Lean Enterprise

A Synergistic Approach to Minimizing Waste

D0921666

Also available from ASQ Quality Press:

Six Sigma for the Shop Floor: A Pocket Guide
Roderick A. Munro

Six Sigma Project Management: A Pocket Guide
Jeffrey N. Lowenthal

Goldratt's Theory of Constraints: A Systems Approach to Continuous Improvement
H. William Dettmer

Customer Centered Six Sigma: Linking Customers, Process Improvement, and Financial Results
Earl Naumann and Steven H. Hoisington

ISO 9001:2000 Explained, Second Edition
Charles A. Cianfrani, Joseph J. Tsiakals, and John E. (Jack) West

Quality Audits for ISO 9001:2000: Making Compliance Value-Added
Timothy O'Hanlon

ISO Lesson Guide 2000: Pocket Guide to Q9001:2000, Second Edition
Dennis Arter and J. P. Russell

ISO 9000:2000 for Small and Medium Businesses
Herb Monnich

The Practical Guide to People-Friendly Documentation
Adrienne Escoe

The Certified Quality Manager Handbook, Second Edition
Duke Okes and Russell T. Westcott, editors

To request a complimentary catalog of ASQ Quality Press publications, call 800-248-1946, or visit our Web site at http://qualitypress.asq.org .

Lean Enterprise

A Synergistic Approach to Minimizing Waste

William A. Levinson and
Raymond A. Rerick

ASQ Quality Press
Milwaukee, Wisconsin

Lean Enterprise: A Synergistic Approach to Minimizing Waste
William A. Levinson & Raymond A. Rerick

Library of Congress Cataloging-in-Publication Data

Levinson, William A., 1957–
   Lean enterprise : a synergistic approach to minimizing waste / William
A. Levinson and Raymond A. Rerick.
       p. cm.
   Includes bibliographical references and index.
   ISBN 0-87389-532-0
   1. Production management. 2. Costs, Industrial. I. Rerick, Raymond
A., 1962– II. Title.
   TS155 .L367 2002
   658.5—dc21   2002004415

Note: As used in this document, the term "ISO 9000:2000" and all derivatives refer to the ANSI/ISO/ASQ Q9000-2000 series of documents. All quotations come from the American National Standard adoptions of these International Standards.

10 9 8 7 6 5 4 3 2 1

ISBN 0-87389-532-0

Acquisitions Editor: Annemieke Koudstaal
Project Editor: Craig S. Powell
Production Administrator: Gretchen Trautman
Special Marketing Representative: Denise M. Cawley

ASQ Mission: The American Society for Quality advances individual, organizational, and community excellence worldwide through learning, quality improvement, and knowledge exchange.

Attention Bookstores, Wholesalers, Schools, and Corporations: ASQ Quality Press books, videotapes, audiotapes, and software are available at quantity discounts with bulk purchases for business, educational, or instructional use. For information, please contact ASQ Quality Press at 800-248-1946, or write to ASQ Quality Press, P.O. Box 3005, Milwaukee, WI 53201-3005.

To place orders or to request a free copy of the ASQ Quality Press Publications Catalog, including ASQ membership information, call 800-248-1946. Visit our Web site at www.asq.org or http://qualitypress.asq.org .

Printed in the United States of America

 Printed on acid-free paper

American Society for Quality

Quality Press
600 N. Plankinton Avenue
Milwaukee, Wisconsin 53203
Call toll free 800-248-1946
Fax 414-272-1734
www.asq.org
http://qualitypress.asq.org
http://standardsgroup.asq.org
E-mail: authors@asq.org

Table of Contents

Preface: Lean Enterprise— A Synergistic Approach

Some day an intelligent nation will awake to the fact that by scientifically studying the motions in its trades it will obtain the industrial supremacy of the world. We hope that that nation will be the United States. . . . Certain it is, that if we do not some other people will, and our boasted progress and supremacy will then be but a memory.

—Robert Thurston Kent, editor of
Industrial Engineering; introduction to
Frank B. Gilbreth's *Motion Study* (1911)

American manufacturers followed this advice, as did Japanese manufacturers decades later. Scientific management, a system that evolved into what we now call lean manufacturing, was directly responsible for the United States' ascendancy over the British Empire—an empire whose own foundations rested in manufacturing prowess—early in the 20th century. During the 1910s the Ford Motor Company and the host of industries that grew to support it made the United States into the wealthiest and most powerful nation on earth. Just-in-time (JIT) and lean manufacturing enabled Henry Ford to write (1930, 11), "Our problem has always been to keep profits down and not up."[1] Japan adopted lean manufacturing and JIT during the 1950s and grew into the second largest economy on earth. The label "Made in Japan," once a proverbial expression for cheap price and poor quality, now commands premium prices and suggests high quality and reliability.

This book is about achieving similar results today through lean manufacturing and lean enterprise. Lean manufacturing is a *system of synergistic and mutually supporting techniques and activities for running a manufacturing or a service operation*. The techniques and activities differ according to the application at hand but they have the same underlying principle: *the elimination of all non-value-adding activities and waste from the business*. Lean enterprise extends this concept through the entire value stream or supply chain. The leanest factory cannot achieve its full potential if it has to work with non-lean suppliers and subcontractors.

Just-in-time is an important element of the lean toolkit, and there are different ways to implement it. Eliyahu Goldratt's theory of constraints (TOC) and synchronous flow manufacturing (SFM) improve on the original Ford system and traditional *kanban* systems. Chapters five and seven in this book on TOC and SFM respectively, apply the extensive experience of Fairchild Semiconductor's Mountaintop, Pennsylvania plant.

This book also shows that lean and JIT are American inventions; the next section provides a brief overview of their history. This is not merely (or even primarily) an interesting historical discussion, it is an extra tool for *change management* and organizational transformation in American workplaces. Change agents in all countries can, meanwhile, use the unquestioned success of lean manufacturing in the United States and then in Japan as a selling point for their own lean and JIT systems.

HENRY FORD TAUGHT JAPAN
HOW TO MAKE CARS

The Japanese, who imported lean manufacturing from the United States during the early 20th century, did not capture a large part of our automobile market by discovering how to make better and cheaper cars. They captured it because Henry Ford (or his books) showed them how to make better and cheaper cars, and then we forgot how to ourselves. Schonberger (1982, 12) shows, in fact, that Japan even adopted Ford's "Any color you want as long as it's black" approach to gain market share:

> They [Japanese firms] got to be giants not by catering to consumer whims but by producing a few models very well, often in market segments that were being ignored by other companies. Low-cost, high-quality production leads to growth in market share (Henry Ford's credo).

The Ford Motor Company introduced just-in-time manufacturing during the first part of the 20th century, and Henry Ford described explicitly the merits of continuous flow and inventory reduction. "While the JIT *concept* (if not the application) is natural in the flow industries, it took Henry Ford and his lieutenants to get JIT worked out in discrete goods manufacturing" (Schonberger 1986, 7). Toyota industrialist Taiichi Ohno[2] credited Ford (and the American supermarket) with the idea.[3] In fact, Norman Bodek, president of Productivity Inc., wrote:

> I was first introduced to the concepts of just-in-time (JIT) and the Toyota production system in 1980. Subsequently I had the opportunity to witness its actual application at Toyota on one of our numerous Japanese study missions. There I met Mr. Taiichi Ohno, the system's creator. When bombarded with questions from our group on what inspired his thinking, he just laughed and said he learned it all from Henry Ford's book (Ford 1926, vii).

All the basic principles of lean manufacturing, as described in Womack and Jones' (1996) *Lean Thinking*, appear in Ford's *My Life and Work* (1922), *Today and Tomorrow* (1926), and *Moving Forward* (1930). These books also describe all the quality and productivity improvement techniques that Japan made famous, including *kaizen* (continuous improvement), *poka-yoke* (error-proofing), *muda* (waste) and *muri* (strain) reduction, and even elements of 5S-CANDO (Levinson 2002).

Ford's Value Today

Earlier scientific management practitioners introduced some lean manufacturing techniques but Henry Ford was the first industrialist to weave them into a comprehensive, synergistic, and mutually supporting system. The complete story (Levinson 2002) has been assembled from the three books by Ford and Samuel Crowther, and many other references. This book's principal focus is lean enterprise plus Dr. Eliyahu Goldratt's theory of constraints and the related synchronous flow manufacturing production control system, and it presents only a brief overview of the Ford story. It does, however, use many excerpts and examples from Ford because:

1. They are often the clearest and sharpest examples and explanations available. Ford, who began his career as a mechanic, lacked an extensive formal education. He was a hands-on person and he wrote from (and to) the very practical perspective of the shop floor—what Masaaki Imai calls *gemba*, the "real place." He explained both lean techniques and the principles behind them in concise, down-to-earth, and easily-understandable terms.

   - The first chapter will show, for example, that *My Life and Work* summarizes all the basic principles of just-in-time manufacturing, plus its transportation considerations, in one paragraph.

2. The transition to lean enterprise is more than the implementation of physical techniques, it requires cultural transformation and organizational buy-in. Change agents may face questions like, "What will lean manufacturing do for us?" Success sells ideas as well as products, and Ford's success was both unprecedented and unquestionable.

- Ford was the world's first self-made billionaire and his enterprises transformed the entire world. We are very fortunate that he left to posterity some books that described how he did it.

- Microsoft CEO Bill Gates is the only other entrepreneur in history to do something of comparable magnitude in terms of both personal success and historical significance. He did so by making computers everyday tools for the middle class, which is what Ford did with cars.

The unquestionable results of lean manufacturing are a selling point for their adoption in any country. Lean manufacturing's "Made in the U.S.A." label is an additional change management asset in the United States. History thus becomes a valuable change management tool.

OVERVIEW OF THE BOOK

The first five chapters of this volume focus on lean manufacturing. Henry Ford offers an outstanding description of lean enterprise in chapter 1 and of JIT in chapter 2. Chapter 4 covers specific techniques and programs for putting lean techniques to work. Remember that most of these techniques are synergistic and mutually supporting, not stand-alone actions that will deliver results by themselves.

Chapter 1 provides a working definition of lean enterprise and stresses the importance of synergy between lean techniques and activities. It then provides another valuable change management tool, a discussion of the importance of manufacturing to national security and prosperity.

Chapter 2 covers the invention of JIT and lean manufacturing. Henry Ford defined the benefits of inventory reduction explicitly. He applied it not only in the factory but also in his supply chain.

Chapter 3 covers the requirements for change management, the transformation of the organizational culture. This requires management commitment, job security, and abolition of restrictive job classifications. Lean thinking applies to the entire organization. It is not a euphemism for downsizing, it means reassigning people and resources from useless work to value-adding work. Layoffs lead to *soldiering*,[4] or efforts by workers to limit production and prevent further improvements that will make their jobs unnecessary.

Chapter 4 covers some specific lean manufacturing methods for making jobs more efficient. The chapter provides more detail about the vital concept of *friction*, which the Japanese call *muda* (waste). Waste can conceal itself even in value-adding activities. A goal of lean manufacturing is to discover and remove this waste. Specific examples are given.

The next three chapters cover Eliyahu Goldratt's theory of constraints, its economic aspects, and its application in synchronous flow manufacturing. TOC shows why improvement efforts must focus on the constraint operation. Productivity gains elsewhere are mostly illusory, they do not really increase the factory's capacity. Batch-and-queue operations are undesirable even if they do not affect capacity because they increase job lead times and complicate statistical process control (SPC).

Chapter 5 treats the theory of constraints and its economic aspects. TOC says that no business system can deliver a product or service more rapidly than its slowest step, the constraint. This chapter adds the managerial economic aspects of TOC, including the opportunity costs of lost production at the constraint.[5] It shows why productivity improvement methods like total productive maintenance (TPM) and single-minute exchange of die (SMED) usually yield the most benefit at the constraint.

Chapter 6 discusses the merits of single-unit processing. Batch processing is undesirable for several reasons. It complicates production scheduling (see Goldratt and Cox 1992). Even one batch-and-queue operation increases production lead times in an otherwise smoothly-flowing process. Batch production can degrade quality by introducing an extra layer of variation and it complicates statistical process control. Batch processing can even preclude reliable estimation of a process' capability indices,[6] which in turn rules out verification of a Six Sigma process. Small-lot and single-unit processing give quality and productivity problems fewer places to hide.

Chapter 7 treats synchronous flow manufacturing. It defines the drum-buffer-rope (DBR) production system, which is the mainstay of the production control system at Fairchild Semiconductor's plant in Mountaintop, Pennsylvania. Henry Ford's books leave an open question, especially for readers of Goldratt and Cox's *The Goal*. Ford's assembly line was essentially a balanced production line, as no operation had excess capacity. It was, in fact, designed to run like a clock, and at close to 100 percent utilization. Why didn't huge piles of inventory accumulate? The answer may lie in the beneficial role of task subdivision in *reducing variation in processing times*. This mitigates against the inventory generation that the matchsticks-and-dice exercise in *The Goal* illustrates, and this has valuable implications for today's practitioners.

Chapter 8 covers the vital principles of supply chain management and supplier development, extending lean principles and SFM to the entire chain of suppliers and subcontractors.

Supply chain management treats organizations in the supply chain as trading partners whose success depends on that of the entire supply chain. This lean manufacturing concept follows the chapters on TOC and SFM because it includes extension of SFM to the entire supply chain. The constraint, wherever it is, sets the pace for all the trading partners. Relatively short transportation distances facilitate just-in-time delivery in Japan and most of Europe. Innovative transportation management can, however, help achieve JIT deliveries of small lots even across longer distances such as in the United States and Russia.

The next two chapters cover operations research techniques that support TOC, SFM, and project planning. Chapter 9 discusses production planning in a constrained factory environment. Linear programming (LP) is a very valuable supporting tool for TOC because it identifies the product mixture that delivers the highest profit subject to limited tool capacity and market demand. It can also account for customer requirements, or orders that must be filled even if they do not yield maximum profit. It also focuses capacity improvement and marketing efforts where they will do the most good. It accounts for limits on factory capacity (theory of constraints) plus external constraints and obligations. It allows easy modeling of "what-if" scenarios, such as the effect of elevating a constraint or making a process step unnecessary.

Chapter 10 applies the theory of constraints to program and project management. Fairchild's Mountaintop site used TOC techniques to complete a semiconductor factory in record time: 13 months between beginning of construction and production start-up. The construction of the plant's 8-inch (200 mm) wafer plant is a case study that exemplifies Goldratt's *critical chain* concept. Project planning techniques like program evaluation and review technique (PERT) and critical path method (CPM) seek to identify the constraint, or longest path, in a project. Crashing an activity on the critical path, or paying more money to accelerate it, is similar to elevating the constraint in TOC and SFM.

William A. Levinson,
Levinson Productivity Systems

Raymond A. Rerick
Fairchild Semiconductor

ENDNOTES

1. Charles Sorensen (1956, 166) shows how Ford's minority stockholders made out "in round numbers" when Ford bought them out in the late 1910s. The numbers were very round, they had a lot of zeros after them. A dollar of original Ford Motor Company stock was worth $2500 sixteen years later, for an annualized growth rate of 63 percent, and this does not even count dividend payments. Ford himself became what may have been the world's first self-made billionaire in an era when a billion dollars was an inconceivable fortune.

2. Japanese write the family name first, for example, Ohno Taiichi, Shingo Shigeo.

3. Smith (1998, 175) explains that Taiichi Ohno saw how warehouse (or supplier delivery) personnel checked supermarket shelves for empty space before trying to add more merchandise. The idea is that the supermarket shelf pulls product from suppliers through empty shelf space.

4. This term's origin probably has nothing to do with a perception that soldiers were lazy or that they looked for ways to avoid working. Taylor (1911a, 30) says, ". . . the slow pace which they [workers] adopt, or the loafing or 'soldiering,' marking time, as it is called." Formal parade drills, which had practical as well as ceremonial applications in the 19th century, sometimes required soldiers to *mark time*, or march in place without going anywhere. (The idea was probably to maintain the unit's synchronization so it could perform the next maneuver upon command.) In a workplace, however, behavior similar to marking time—moving without accomplishing anything—would be a way to limit production.

5. Benjamin Franklin's *Poor Richard's Almanac* identified the opportunity cost concept in the 18th century: "He that idly loses 5s. worth of time, loses 5s., and might as prudently throw 5s. into the river. He that loses 5s. not only loses that sum, but all the other advantages that might be made by turning it in dealing, which, by the time a young man becomes old, amounts to a comfortable bag of money."

6. The traditional process capability indices measure the process' ability to meet specifications. C_p is a ratio of the specification width to the process variation. CPL and CPU measure the ability to meet the lower and upper specifications respectively. C_{pk} is the minimum of CPL and CPU.

Several of the figures in this book were prepared with Corel Gallery clip art, CorelDraw, and Corel PhotoPaint.

1

What Is Lean Enterprise?

The terms *lean manufacturing* and *lean enterprise* appear very frequently in business literature. This chapter begins with an overview of what they really are. Henry Ford (1922, 147–48) defined the lean concept in one sentence: "We will not put into our establishment anything that is useless." The *APICS Dictionary* (Cox and Blackstone 1998) defines lean enterprises and lean manufacturing in more detail but the basic idea is the same. The idea is to identify and eliminate non-value-adding activities from every aspect of the business.

A single factory can practice lean manufacturing but a lean enterprise includes the entire supply chain or value stream. This concept is important because, for example, a supplier's or subcontractor's batch-and-queue operation can raise havoc in the leanest factory's production management system. This points to the importance of supply chain management, to which chapter 8 of this book is devoted.

FRICTION

Friction, a basic idea in lean enterprise, is from General Carl von Clausewitz's *On War* (1976). It means, ". . . the force that makes the apparently easy so difficult. . . . countless minor incidents—the kind you can never really foresee—combine to lower the general level of performance, so that one always falls short of the intended goal." Japan calls friction *muda* (waste). *A lean enterprise is one from which friction is absent.* Since friction is never completely absent, there are only degrees of leanness.

This concept is so important and pervasive that it appears in many references. Henry Ford almost quotes Clausewitz on friction:

It is the little things that are hard to see—the awkward little methods of doing things that have grown up and which no one notices. And since manufacturing is solely a matter of detail, these little things develop, when added together, into very big things (Ford 1930, 187).

Dr. Shigeo Shingo echoes this idea:

Unfortunately, real waste lurks in forms that do not look like waste. Only through careful observation and goal orientation can waste be identified. We must always keep in mind that the greatest waste is the waste we don't see (Robinson 1990, 14).

So does Taiichi Ohno:

In reality, however, such waste [waiting, needless motions] is usually hidden, making it difficult to eliminate. . . . To implement the Toyota production system in your own business, there must be a total understanding of waste. Unless all sources of waste are detected and crushed, success will always be just a dream (Ohno 1988, 59).

In addition:

The accumulation of little items, each too trivial to trouble the boss with, is a prime cause of miss-the-market delays (Peters 1987).

The machine that jams sometimes, the tool that must be searched for sometimes,[1] the assembler who does the task the wrong way sometimes, the part that arrives late sometimes, the blueprint that is wrong sometimes, the part that is off the mark sometimes—all of these and many more require costly sets of 'solutions.' They are not true solutions, because they provide ways to live with the problems (Schonberger 1986, 14–15).

Watch the little points. They spell success or failure; they are treacherous; they will knife your profits (The System Company 1911a, 11).

All these people are talking about exactly the same thing: the seemingly minor inefficiencies that people live with or work around. Levinson and Tumbelty (1997) provide a one-sentence definition for frontline workers: *If it's frustrating, a chronic annoyance, or a chronic inefficiency, it's friction.*
Inventory is one form of waste but it is often the result, not the cause, of friction. Standard and Davis (1999) call inventory the flower and not the root of all evil. JIT is about reducing inventory but it cannot work properly

while inventory's root causes remain. Batch-and-queue operations and non-JIT suppliers and subcontractors, for example, make it difficult for even the best JIT system to get rid of inventory. Lean manufacturing eradicates these root causes so it is almost a prerequisite for successful JIT. Correction of the root causes, in fact, often makes the inventory go away with little or no further effort.

The Basic Idea Is to Eliminate Waste

Modern practitioners have, of course, extended and improved Ford's methods but the basic principles remain the same. Waste, including waste of time, is absolutely intolerable. Idle material and work in process—inventory—wastes time. The idea that everything in a business must add value to the product or service is paramount:

> We will not put into our establishment anything that is useless. We will not put up elaborate buildings as monuments to our success. The interest on the investment and the cost of their upkeep only serve to add uselessly to the cost of what is produced—so these monuments of success are apt to end as tombs (Ford 1922, 147–48) (Figure 1.1).

An executive office was once really something to die for . . . . These structures were also full of hieroglyphics ("sacred writings"). *Hieroglyphics* has the same root as *hierarchy,* a characteristic of non-lean companies with pyramidal organizational charts.

Figure 1.1   The world's first monuments of success.

This section has defined a lean enterprise in terms of a single concept: friction. A lean enterprise is one from which friction is absent. Lean enterprise applies to an *entire* value stream or supply chain. Lean manufacturing, a subset, takes place in a single factory. The difference is important because a non-lean supply chain can prevent the leanest factory from achieving its full potential. The next section adds another vital concept, that of synergy.

SYNERGY: SEEING AND USING THE WHOLE ELEPHANT

This book's goal is to present a holistic approach, or *a synergistic and mutually supporting set of activities and techniques.* These activities and techniques all work together as a lean manufacturing *system.* The sidebar on page 5 displays a well-known poem that stresses the difference between a set of isolated tools and a system.

Saxe applies the parable to theological disputes but the same idea applies to management systems. Legs, tusks, sides, ears, a trunk, and a tail are elephant parts but none of them makes up an elephant by itself. Even if all are present they must work together and support one another's activities before one has an elephant. The same applies to lean enterprise (Figure 1.2).

We cannot overemphasize the system concept. Attempts to implement the alphabet soup of *kaizen, kanban,* SMED, 5S-CANDO, Six Sigma, *poka-yoke,* preventive maintenance, and so on on a piecemeal basis will

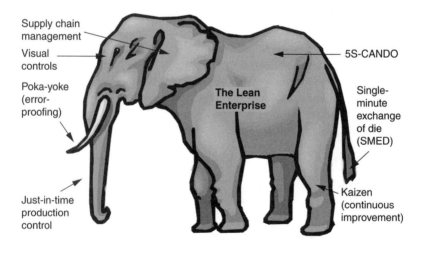

Supply chain management

Visual controls

Poka-yoke (error-proofing)

Just-in-time production control

The Lean Enterprise

5S-CANDO

Single-minute exchange of die (SMED)

Kaizen (continuous improvement)

Figure 1.2   Synergy and lean enterprise.

John Godfrey Saxe (1816–1887), *The Blind Men and the Elephant*

It was six men of Indostan
To learning much inclined,
Who went to see the Elephant
Though all of them were blind,
That each by observation
Might satisfy his mind.

The First approached the
    Elephant
And, happening to fall
Against his broad and sturdy side,
At once began to bawl:
"God bless me, but the Elephant
Is very like a wall!"

The Second, feeling the tusk,
Cried, "Ho! what have we here
So very round and smooth
    and sharp?
To me 'tis very clear
This wonder of an Elephant
Is very like a spear!"

The Third approached
    the animal
And, happening to take
The squirming trunk within
    his hands,
Thus boldly up he spake:
"I see," quoth he, "The Elephant
Is very like a snake!"

The Fourth reached out an
    eager hand,
And felt about the knee:
"What most the wondrous beast
    is like
Is very plain," quoth he;
"Tis clear enough the Elephant
Is very like a tree!"

The Fifth, who chanced to touch
    the ear,
Said, "Even the blindest man
Can tell what this resembles most;
Deny the fact who can:
This marvel of an elephant
Is very like a fan!"

The Sixth no sooner had begun
About the beast to grope
Than, seizing on the swinging tail
That fell within his scope,
"I see," quoth he, "the Elephant
Is very like a rope!"

And so these men of Indostan
Disputed loud and long,
Each in his own opinion
Exceeding stiff and strong.
Though each was partly in
    the right,
They all were in the wrong!

Moral:
So oft in theologic wars,
The disputants, I ween,
Rail on in utter ignorance
Of what each other mean,
And prate about an Elephant
Not one of them has seen!

probably lead to frustration and disappointment. An organization that tries to focus on a dozen or more different initiatives will end up focusing on nothing. Frederick II ("The Great") of Prussia introduced this concept as, "One who tries to defend everything ends up defending nothing."

Implementation becomes far easier by treating the different techniques and activities as aspects of the same system. The basic idea of lean manufacturing is to suppress all forms of waste. Don't look at the alphabet soup as a selection of different initiatives and programs, each of which can become the "program of the month" all too easily. Instead, treat every technique in this book (and some that aren't) as tools in a single toolbox—the lean manufacturing toolbox. Select the right tool for the job at hand. Two or more tools can often work together to yield results that none can achieve alone; this is synergy.

This chapter's next section shows the importance of manufacturing capability, which lean manufacturing preserves and develops, to national prosperity and security. Readers can use the material in this section to:

1. Promote manufacturing organizations' morale by underscoring the role of manufacturing in the nation's welfare. Remove the perception of marketing and finance as glamorous upscale occupations that are preferable to so-called "dirty" manufacturing jobs. Marketing would have nothing to sell and finance would have no money to manage without manufacturing.

2. Educate the lay public about manufacturing's role as the foundation of a high standard of living. The public's perception that service sector jobs are preferable replacements for "dirty" manufacturing jobs is extremely dangerous to the nation's welfare and military security. "Manufacturing is the only sector that's not doing well" shows exactly why the stock market has been performing very poorly for almost the past two years (through December 2001).

MANUFACTURING: A VITAL ISSUE

You can raise all the wheat, corn, cattle, and hogs you want; you can mine all the raw materials you want . . . but only manufacturing can guarantee good jobs and the good life. Only manufacturing can guarantee the creation of personal and national wealth. Only manufacturing can guarantee international strength. . . . If we were pagans, we would put manufacturing on a pedestal, build temples to it, and worship it.

—Albert W. Moore (1996)

The actual wealth of the nation is in what it takes from the ground in the shape of crops or minerals plus the value added to these products by processes of manufacture. If by reducing the number of motions in any of these processes we can increase many fold the output of the worker, we have increased by that amount the wealth of the world; we have taken a long step in bringing the cost of living to a point where it will no longer be a burden to all but the very wealthy; and we have benefited mankind in untold ways.

—Robert Thurston Kent, in Gilbreth (1911)

The statement by Kent covers the roles of manufacturing, and particularly lean manufacturing, in creating wealth. Manufacturing is so important that it was literally something, to use a modern expression, "to die for." Monopolization of manufacturing capability through laws or tariffs played a direct role in starting the War of Independence and the American Civil War.

Manufacturing is the foundation of national prosperity and national security. It is the only activity that adds value—and lean manufacturing is all about adding value—to raw materials that come from extractive industries like mining, lumbering, and agriculture.

The foundations of society are the men[2] and means to *grow* things, to *make* things, and to *carry* things. As long as agriculture, manufacture, and transportation survive, the world can survive any economic or social change (Ford 1922, 6–7).

Goldratt and Fox (1986) add:

Manufacturing has been the major wealth generator of our industrialized world. This ability to generate wealth has made our standard of living the envy of the rest of the world. If we continue to lose our manufacturing base, and we are losing it rapidly, we and everyone else will certainly live less well.

Even the Soviet Union knew this, as shown by its hammer-and-sickle flag. The Communists' goal was to acquire the means of production, not banks and other financial and service enterprises. They were right about what was important but they couldn't run it once they had their hands on it. This is why abundant food and consumer goods could often be found in five-year plans (government-issued plans for the country's economic future) but nowhere else.

The Influence of Manufacturing on History

Quarrels over manufacturing capability—specifically protectionist laws and tariffs—were major causes of the War of Independence and the American Civil War. Industrialization played a major role in ending slavery in Great Britain and the northern United States. Superior manufacturing capability was decisive in the Union's victory in the Civil War, Allied victory in the Second World War, and the relatively nonviolent overthrow of the Soviet Union in what was effectively the "Third World War."

This discussion should underscore why the United States must stem the loss of its manufacturing capability if it is to continue as an affluent world power. Lean manufacturing is the way to do this. We provide this information for its historical interest only as a secondary consideration. Lean practitioners and change agents can use it to sell the lean enterprise concept to organizations and gain buy-in by workers at all levels.

Begin with Spain's discovery of the New World and its abundant gold in 1492. A typical "bean counter" who equates money with genuine wealth might conclude that this windfall should have made Spain the most powerful nation on earth. Any such illusion lasted for less than a century. The truth manifested itself with the destruction of the Spanish Armada at Gravelines in 1588; England had become the most powerful nation on earth.

Alfred Thayer Mahan's *The Influence of Sea Power Upon History* (1980, 46) explains that Spain and Portugal (the latter found silver in Brazil) used their so-called wealth to buy manufactured goods from their rivals England and Holland. This developed the latter countries' manufacturing capabilities and shipbuilding industries—the raw materials and finished goods required transportation—at the former countries' expense. Mahan summarizes, "The tendency to trade, *including of necessity the production of something to trade with*, is the national characteristic most important to the development of sea power."

The British now had a good thing and they knew it. Parliament enacted protectionist legislation like the Navigation Acts, which gave English ships a monopoly on all transportation to England and its colonies. Other legislation forbade the colonies to manufacture anything; they were to export raw materials to England and import all their manufactured goods. This was, along with "taxation without representation," a major cause of the War of Independence. English human rights were actually ahead of most other European nations' prevailing standards, but the country also forbade the emigration of skilled textile workers. England prohibited many manufacturing activities in its colonies and furthermore:

> Certain parts of England had a monopoly on certain trade and some manufactures and importations. When British merchantmen brought back from the Indian port of Calicut a beautiful cotton

cloth which later became known as calico, the British wool merchants were thrown into a panic. The cloth was so beautiful and so cheap that they foresaw the ruin of their own industry. Parliament readily complied with their demand that the importation of this textile be banned and the manufacture of cotton be prohibited (Crow 1943, 26–27).

Once the American colonies achieved their independence they went to enormous lengths to obtain and even steal British manufacturing technology. One American obtained a textile machine, cut it apart, shipped it to France as "plate glass," shipped it from France to America, and then reassembled it.

The Industrial Revolution, meanwhile, destroyed slavery in Great Britain and probably would have ended it in the United States even if the Civil War had not taken place. Industrialization is the death of slavery because it reduces the need for low-cost unskilled labor. As a country industrializes, its working class begins to oppose slavery because slaves can displace paid workers. Henry Ford described why subsistence wages (of which slavery is the ultimate extension) retard economic growth: If the workers can't afford to buy what they make, their employer has far fewer customers.

The difference between Northern and Southern industrial capability was, in fact, a principal cause of the American Civil War. Northern factory owners wanted tariffs to protect them against English manufactured goods. The agrarian South sold much of its cotton to England's Lancashire textile mills, and the South wanted to import English manufactures. The South found the protectionist tariffs ruinous, and this led to secession and war. The North's superior manufacturing capability was then decisive in wearing down the Confederacy and forcing its surrender.

Admiral Yamamoto warned Japan that it could not win a war with the United States. American industrial power was the "sleeping giant" that, once awakened, crushed the Axis during the Second World War. Taiichi Ohno might have agreed with Yamamoto after learning in 1937 that German workers were three times as productive as Japanese workers, and Americans were three times as productive as Germans (Ohno 1988, 3). The caption of "The Fighting Pacifist," a cartoon in the 1918 *London Times*, should have served as an even earlier warning to Japan's warlords: "Henry Ford is the most powerful individual enemy the Kaiser has." The cartoon shows Henry Ford throwing a steady stream of tanks, trucks, ships, airplanes, shells, and money (in the form of support for the war effort) at the Kaiser (Alvarado and Alvarado 2001, 74).

Lean manufacturing was why American workers were then nine times as productive as Japanese workers. Charles Sorensen (1956, 273) leaves no doubt as to what happened: "The seeds of [Allied] victory in 1945 were sown in 1908 in the Piquette Avenue plant of Ford Motor Company when we experimented with a moving assembly line." He adds as a further example (1956, 4), ". . . a pencil sketch I made in a California hotel room was the beginning of the mile-long Willow Run plant which ultimately turned out one B-24 bomber an hour during World War II." There was no way for the Axis to match that kind of production.

The United States won the 'Third World War' by confronting the Soviet Union with an arms race that its economy could not support. These lessons underscore the role of manufacturing in national security.

The Problem: Manufacturing's Decline

Marciano (1999) cites historian Ellsworth Grant in explaining why Hartford, Connecticut, which could once claim to be the wealthiest city in the United States, is now among the ten poorest. From the Civil War to the 1929 Depression, the Colt Armory, Pratt & Whitney, and 60 or so other factories were the foundation of the city's economy. Then "service-oriented jobs began to outstrip those in manufacturing . . . 'I think 1960 was really a watershed year for Hartford,' says Grant, 'because that was the year the service industry took over as the major employer, and it was also the year Constitution Plaza was built. And that decimated the vitality of downtown Hartford.'"

In other words, people can't build a viable economy by selling each other hamburgers or even Hartford's other famous products, insurance and banking services. "Would you like fries with that insurance policy?"

Even more alarming is the fact that less than a third of the United States' economy produces wealth. Table 1.1 shows the breakdown of the 1999 gross domestic product (GDP; percentages do not add up to 100 due to rounding).[3] Transportation and utilities are treated as necessary to manufacturing.

Some services are necessary. Healthcare, for example, comprises 5.5 percent of the GDP. Part of the Federal government figure includes the guns that protect our butter as well as our freedom. Wholesale and retail trade, on the other hand, equal manufacturing. This suggests that 50 percent of the price of manufactured goods consists of waste: the intermediaries' overhead, inventory carrying costs, floor space, and profit. The finance, insurance, and real estate figure shows that shuffling wealth and taking a cut from each transaction is a bigger part of the economy than manufacturing.

In contrast, industry is 35 percent of China's gross domestic product and manufacturing employs 24 percent of the Chinese workforce. China's manufacturing growth rate is nine percent and this far exceeds the United States' manufacturing growth rate. Chinese manufacturing totals $1.68 trillion, almost three-quarters of the United States' $2.3 trillion (Briscoe 2001).

Table 1.1 Wealth-producing and non-wealth-producing segments of U.S. GDP.

Activity	Percent
Extractive industries and agriculture	2.5%
Construction	4.5%
Manufacturing	16.1%
Transportation and public utilities	8.4%
Wealth-producing activities	**31.5%**
Wholesale trade	6.9%
Retail trade	9.2%
Finance, insurance, and real estate	19.3%
Services	21.4%
Government	12.5%
Activities that do not produce wealth	**69.3%**

The United States' standard of living is much higher because only one fifth as many workers share in this real wealth. (The same is true of most European nations, Japan, Taiwan, South Korea, and Canada; there is more production per worker.) The problem is that more than two-thirds of the United States' supposedly massive gross domestic product is mostly an illusion. *Manufacturing comprises less than one sixth.* This is extremely dangerous because military power is a direct function of net manufacturing capability.

This is not to say that the United States and other developed nations should resent or oppose industrialization in other countries. Henry Ford believed that this trend would promote world peace by removing poverty, a major root cause of war. Industrialization was a major factor in ending slavery and then the slave-like working conditions of the 19th century. It can also, by increasing other nations' wealth, create more customers for American goods.

The U.S. must, however, avoid the further loss of its own manufacturing capability to other countries. The steady replacement of American manufacturing jobs by so-called service jobs and dot-com companies is a clear and present danger to the country's prosperity and military security. Loss of manufacturing capability is a fatal disease; lean manufacturing is the cure.

The Answer: Lean Manufacturing

Henry R. Towne, past president of the American Society of Mechanical Engineers (ASME), wrote that lean (or efficient) manufacturing is the answer to the hemorrhage of manufacturing jobs—then from the United

States, now from places like Japan and the European Economic Community as well—to countries with cheap labor:

> We are justly proud of the high wage rates which prevail throughout our country, and jealous of any interference with them by the products of the cheaper labor of other countries. To maintain this condition, to strengthen our control of home markets, and, above all, to broaden our opportunities in foreign markets where we must compete with the products of other industrial nations, we should welcome and encourage every influence tending to increase the efficiency of our productive processes (Foreword to Taylor 1911a, 10).

Nothing has happened during the past 90 years to change this. In 1984, for example, a Toyota four-cylinder engine required less than 20 percent of the labor as a similar engine from Chrysler or Ford (Womack and Jones 1996, 59). Toyko paid $11.35 (1984 money) for one hour. The Americans, at the industry average of $12.50/hour, were paid $62.50 for five hours. Lean manufacturing makes every worker more productive. Taylor (1911) cites worker productivity as the principal difference between wealthy and poor nations.

This section has shown just how important manufacturing is to the wealth and security of any nation. Lean manufacturing protects and nurtures manufacturing capability. The reader may use this material to promote lean cultural change in his or her organization, and also to help influence public policy and public awareness.

The next chapter will describe the origins of JIT and lean manufacturing. The methods and especially the principles are still applicable to modern enterprises. Chapter 2 also describes the role of change management in adopting lean manufacturing and introduces the Toyota production system.

ENDNOTES

1. 5S-CANDO includes arranging the workplace so as to provide a specific place for every tool.
2. Don't worry about the gender-specific word "men" in the older references (Ford, Taylor, Basset, The System Company). These books were written in an era when workforces were predominantly male. Although Ford embraced the now-unacceptable idea that a woman whose husband was employed should not work, he was far ahead of the National Organization for Women in saying that men and women should receive equal pay for equal work.
3. Bureau of Economic Analysis, Industry Accounts Data, gross domestic product by industry. http://www.bea.doc.gov/bea/dn2/gposhr.htm#1993–99 as of 5/04/01.

2

The Birth of JIT and Lean Manufacturing

Schonberger (1982, 205) says Toyota's JIT production system ". . . may be the most important productivity-enhancing management innovation since Taylor's scientific management at the turn of the [20th] century." Toyota did not, however, invent JIT, the underlying principles of which Henry Ford described in one paragraph:

> We have found in buying materials that it is not worth while to buy for other than immediate needs. We buy only enough to fit into the plan of production, taking into consideration the state of transportation at the time. If transportation were perfect and an even flow of materials could be assured, it would not be necessary to carry any stock whatsoever. The carloads of raw materials would arrive on schedule and in the planned order and amounts, and go from the railway cars into production. That would save a great deal of money, for it would give a very rapid turnover and thus decrease the amount of money tied up in materials. With bad transportation one has to carry larger stocks (Ford 1922, 143).

The key points are:

1. Materials arrive only when, and exactly when, the production line needs them

2. Materials go not only from "dock to stock" but from dock to factory floor

3. JIT depends on reliable transportation and logistics

4. Inventory ties up capital; getting rid of inventory frees capital

5. Cycle time reduction also frees capital

1. *Materials arrive only when, and exactly when, the production line needs them.* In Goldratt and Cox's *The Goal* (1992, 245–46) UniCo gets a contract for 1000 Model 12s for delivery to Bucky Burnside's company. Alex Rogo, UniCo's plant manager, says, ". . . there is no way we can deliver the full thousand units in two weeks. But we can ship 250 per week to them for four weeks." The sales manager then discovers, "They even like the smaller shipments *better* than getting all thousand units at once!"

2. *Materials go not only from "dock to stock" but from dock to factory floor.* Ford (1926, 118–19) describes the ideal progress of iron ore from a Great Lakes freighter to production: "Ten minutes after the boat is docked, its cargo will be moving toward the High Line and become part of a charge for the blast furnace." In Ford's Chicago plant, incoming materials traveled no more than 20 feet (6.1 meters) from the railroad freight car to the first conveyor. There were no warehouses or storage yards. There were apparently no incoming quality inspections.[1] Dock to stock, or dock to factory, therefore requires absolute confidence in the quality of the incoming materials.

3. *JIT depends on reliable transportation and logistics.* JIT also demands absolute reliability from the logistics system. Ford (1926, 117) identified the following issues:

- The material in transit, which Ford called the "float," is really inventory.

- The shortage of any component will hold up production.

- Material must be ordered and delivered on time, and exactly on time. Ford's system for achieving this was particularly impressive. Although computers did not exist, a shipment's status could be pinpointed to within an hour. Headquarters would know if a rail car was more than an hour overdue.

Heizer and Render (1991, 573) show that the "float" is still a problem. "A GM executive estimates that at any time more than half of the company's inventory is on trucks and trains."

The Internet, and business-to-business computer networks, offer improved logistics capabilities today. Chapter 8 on supply chain management discusses *freight management services* (FMSs) and *third party logistics* (3PL) systems.

Weyerhaeuser Corporation, a major pulp and paper manufacturer, uses its Intranet to manage shipments to a customer's factory. The system links the customer's sales information to Weyerhaeuser's Valley Forge Fine Paper Company's master production schedule. The customer's factory sets the pace for the supplier's factory. As the customer uses the Weyerhaeuser products, the Valley Forge mill schedules more. The system reduces the customer's inventory by up to 40%, with the benefits that Ford has already described (Richards 1996). A senior Weyerhaeuser manager says, "We're just-in-timing every aspect of the manufacturing, ordering, and delivery process." The company expects the system to ". . . double the unit's productivity without adding any additional staff, allowing it to undercut competitor's prices and grab additional market share" (Richards 1996).

4. *Inventory ties up capital, getting rid of inventory frees capital.* Readers of Goldratt and Cox's (1992) *The Goal* will appreciate the following achievement:

> The extension of our business since 1921 has been very great, yet, in a way, all this great expansion has been paid for out of money which, under our old methods, would have lain idle in piles of iron, steel, coal, or in finished automobiles stored in warehouses. We do not own or use a single warehouse (Ford 1926, 112).

The practice of taking over companies by using their own cash has been around for a while. Womack and Jones (1996, 147) show how Wiremold takes over companies by using their own inventory. "Each time Wiremold's vacuum sucks up a batch-and-queue producer it spits out enough cash to buy the next batch-and-queue producer!" Wiremold's procedure is similar to Ford's: convert inventory to cash.

Standard and Davis (1999) raise the important point that inventory is not the root of evil but rather its flower: a symptom and not an underlying cause. Underlying problems include variation in production rates, large batch sizes, and incentives to make unnecessary parts to keep tool efficiencies high. High inventories of finished goods may exist to allow rapid fulfillment of customer orders. The underlying cause for this inventory could be a long production cycle time that prevents making-to-order rapidly enough to satisfy the customers. This inventory cannot therefore be eliminated without first reducing the cycle time.

5. *Cycle time reduction also frees capital:*

> We had been carrying an inventory of around $60,000,000 to insure uninterrupted production. Cutting down the time one third released $20,000,000, or $1,200,000 a year in interest. Counting the finished inventory, we saved approximately $8,000,000 more—that is, we were able to release $28,000,000 in capital and save the interest on that sum (Ford 1922, 175).

This section has shown that JIT and the underlying concepts originated not in Japan but in the United States. Change agents can use this information to sell the lean enterprise concept in American workplaces. Change agents can use the success of the early Ford Motor Company and of Japanese companies that embraced lean enterprise concepts to sell these ideas to any country's workforce. The next section introduces the Ford system's successor, and a well-known model for lean manufacturing, the Toyota production system. Key elements include pull production control (through *kanban*), quality, autonomation, and flow.

THE TOYOTA PRODUCTION SYSTEM

Taiichi Ohno used the Ford system as the foundation for the Toyota production system. Where Ford produced one product ("You can have it in any color you want as long as it's black") to hold costs down, Ohno sought to produce small lots of many products with equal efficiency. Other key principles such as removing all forms of waste and holding minimal inventory were the same as Ford's. Ohno (1988, 4) defines the basis of the Toyota production system as the absolute elimination of waste. The two pillars needed to support the system are:

- Just-in-time

- Autonomation, or automation with a human touch

Pull

The Toyota system, or JIT, is emphatically a pull system. Production control does not push work into the system at the first operation; downstream processes call for work as they complete the jobs they have.

Quality

JIT also requires reliable quality. If a downstream process calls for parts and the upstream process makes defective ones, the downstream process will run out of work. The old contingency plan was to have a lot of inventory at every operation. This increases carrying costs and the risk of inventory obsolescence. Defective inventory, and the defects' root causes, may not be discovered for quite some time.

Synchronous flow manufacturing, which this book will examine in chapter 7, keeps an inventory buffer at the constraint, or lowest-capacity operation. This prevents throughput loss even if something does go wrong.

Autonomation (Jidoka)

SFM protects the factory from throughput losses when problems occur; autonomation *(jidoka)* prevents the problems. It means imparting human intelligence to machines. Sakichi Toyoda (1867–1930), the founder of the Toyota Motor Company, invented an automatic weaving machine that stopped if any of the threads broke. An *autonomated* machine is therefore one that can distinguish between normal and abnormal conditions. It does not require continuous operator attention and a single worker can therefore handle several machines.

Automatic detection of tool wear is an example of autonomation. One can, of course, stop the machine and inspect the cutting tool periodically. This slows production (unless the machine can continue to work with another cutting tool, which is quite possible under SMED) and requires human attention. Wearout of a drill or tap is detectable by monitoring the torque, which begins to rise at the end of the tool's life (Kalpakjian 1984, 510). This is an example of autonomation because it enables the tool to detect problems automatically.

An experienced machinist might detect tool wearout by noticing that the machine "just doesn't sound right." Acoustic emission equips the machine itself to know when it doesn't sound right. A piezoelectric transducer on the toolholder detects acoustic emissions from the stress waves that the cutting operation generates. The root-mean square (RMS) of the acoustic emission signal increases with tool wear. The technique is applicable to turning and drilling (Kalpakjian 1984, 498). Wayne Smith (1998, 151) adds the technique of testing lubrication oil for metal content, presumably to detect wear on machine parts.

Shingo (1986, 241) describes a heat-treatment process in which stoppage of the mesh conveyor belt resulted in defective products. (The product presumably stayed in the furnace too long.) The factory installed a poka-yoke (error-proofing) device that detects stoppage of a mesh conveyor belt, notifies factory personnel, and shuts down the drive to prevent damage to the belt.

An autonomated workstation can announce trouble through a light panel (andon) and/or an audible signal. The andon can also announce a normal condition, such as the completion of a job.

If an abnormal condition stops an autonomated workstation the operator shouldn't simply fix and restart it. It could be a chronic problem that people must keep correcting—or *friction.* Ohno references a Japanese saying about hiding an offensive-smelling object by covering it up. Air Force general Curtis LeMay said, "We should stop swatting flies and go after the manure pile." The idea is the same: fix the problem's root cause and keep it from coming back (Figure 2.1). Chapter 4 on lean manufacturing techniques includes a section on Ford Motor Company's Team-Oriented Problem Solving, 8 Disciplines (TOPS-8D), a systematic approach for doing this.

Flow

Ohno recognized the drawbacks of the process-oriented or job-shop factory layout, in which departments are organized by machine type. The tendency in such factories is to process and transport batches of parts, thus creating inventory and promoting non-value-adding transportation costs.

Henry Ford had already identified the foundation of cellular manufacturing and the work cell: organize departments by part type and equip each department with all the machines necessary to make the part. Ohno writes, "As an experiment, I arranged the various machines in the sequence of machining process. This was a radical change from the conventional system in which a large quantity of the same part was machined in one process and then forwarded to the next process" (1988, 11).

Production leveling is related to flow. If demand for a part is 1000 per month, production should be 40 per day if the month contains 25 working

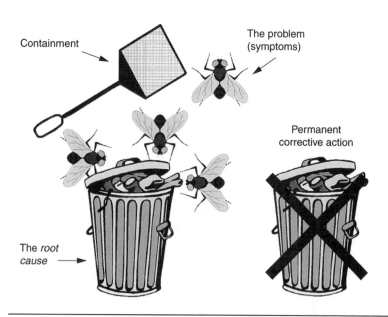

Figure 2.1   Containment versus permanent corrective action.

days. Furthermore, if the work day is 480 minutes, the production rate should be one unit every 12 minutes. This is the essence of *takt time*:

$$\text{takt time} = \frac{\text{available production time}}{\text{demand for the production period}} = \frac{\text{time}}{\text{unit}}.$$

Production leveling helps assure smooth flow at all operations, as opposed to requiring awkward peak-and-valley production levels. We will see that variation in per-unit processing times is a major cause of inventory. The matchsticks-and-dice exercise in Goldratt and Cox's *The Goal* is a simple illustration of this concept.

The well-known kanban card is the downstream process-driven production order in the Toyota production system. Its purpose is to assure flow instead of inventory accumulation. This book will cover kanban and other pull systems in chapter 6.

This chapter has covered the American origins of lean manufacturing and the lean enterprise and an overview of the Toyota production system, a further evolution of American lean enterprise methods. Change agents can use this information to promote cultural transformation in their organizations, which is the subject of the next chapter.

ENDNOTE

1. Ford's assembly lines had plenty of "snap," or go/no-go, gages that culled out nonconforming pieces automatically. Where possible, sorting took place as the parts moved along; the pieces did not stop in front of a quality inspector.

3

Lean Cultural Transformation

L ean enterprise is not merely a set of mutually supporting techniques, it's a change in the entire organization's culture and thought processes. This chapter explores the organizational and cultural changes that are prerequisites for a successful lean enterprise.

We cannot overemphasize the point that managers and engineers cannot create a truly lean enterprise by themselves. H. Ross Perot once said, "When they see a snake at GM, they form a committee to study snakes. Then they bring in a snake consultant. Then they hire a snaking manager. At EDS, when we see a snake, the closest EDSer kills it."[1] We want to empower people at all levels of a business to recognize friction (muda) and deal with it on sight. The next section discusses the vital concept of change management, the successful cultural transformation that achieves this goal.

CHANGE MANAGEMENT

Change management means (1) persuading the organization that change is necessary and (2) getting organizational participants to accept or, even better, internalize the changes.

Frederick Winslow Taylor (1911, 1911a) described this need more than 90 years ago. Lathin and Mitchell (2001) cite the need for sociotechnical system (STS) integration, which recognizes that work organizations combine technical and social systems that must support each other; neither will work without the other. Lathin and Mitchell offer a lean implementation planning matrix the concept of which is similar to quality function deployment's interaction matrix. The roof of the quality function deployment (QFD) house of quality shows positive interactions (synergies) and negative interactions (tradeoffs or barriers) between product characteristics. The lean implementation matrix does the same for technological factors versus social factors.

Standard and Davis (1999, 70) add, ". . . the effectiveness [of lean manufacturing techniques] seems to come when propitious cultural values are blended with the practices." Not only are lean manufacturing techniques synergistic and mutually supporting, culture and technology also must support each other.

Change is more effective when it is something that people do, not something management does to them. Under Taylor's original version of scientific management, managers changed the culture by introducing work efficiency methods and rewarding workers for compliance. One approach was to select a few workers, gain their cooperation by offering higher wages, and getting them to adopt the new methods. Other workers then understood that they also could get more pay by accepting the new system. Taylor's approach to a Bethlehem pig-iron handler shows that the method relied heavily on extrinsic motivation or the use of money or similar benefits to gain compliance:

> The task before us, then, narrowed itself down to getting Schmidt to handle 47 tons of pig iron per day and making him glad to do it. This was done as follows. Schmidt was called out from among the gang of pig-iron handlers and talked to somewhat in this way:
>
> "Schmidt, are you a high-priced man?"
>
> "Vell, I don't know vat you mean."
>
> . . . "What I want to find out is whether you are a high-priced man or one of these cheap fellows here. What I want to find out is whether you want to earn $1.85 a day or whether you are satisfied with $1.15, just the same as all those cheap fellows are getting."
>
> "Did I vant $1.85 a day? Vas dot a high-priced man? Vell, yes, I vas a high-priced man."
>
> . . . "Now, hold on, hold on. You know just as well as I do that a high-priced man has to do exactly as he's told from morning till night. You have seen this man here [presumably one of Taylor's assistants] before, haven't you?"
>
> "No, I never saw him."
>
> "Well, if you are a high-priced man, you will do exactly as this man tells you to-morrow, from morning till night. When he tells you to pick up a pig and walk, you pick it up and you walk, and when he tells you to sit down and rest, you sit down. You do that right straight through the day. And what's more, no back talk. Now a high-priced man does just what he's told to do, and no back talk. Do you understand that? When this man tells you to walk,

you walk; when he tells you to sit down, you sit down, and you don't talk back at him. Now you come on to work here to-morrow morning and I'll know before night whether you are really a high-priced man or not" (Taylor 1911, 20–21).

This story is probably the root of the perception that Taylor wanted workers to leave their brains at the factory gate. He emphasized, however, that he was dealing with an unskilled worker whose primary interest was high wages instead of intrinsically-motivating work:

> This seems to be rather rough talk. And indeed it would be if applied to an educated mechanic, or even an intelligent laborer. With a man of the mentally sluggish type of Schmidt it is appropriate and not unkind, since it is effective in fixing his attention on the high wages which he wants and away from what, if it were called to his attention, he probably would consider impossibly hard work (Taylor 1911, 20–21).

A later section on standardization and best practice deployment shows, in fact, that Taylor wanted the frontline workers to suggest ways to make their jobs more efficient. He was therefore well ahead of modern management experts on the idea of worker empowerment. The following reinforces the frontline worker's role in improving productivity under scientific management:

> In another instance, a workman will stop you when you are passing and ask whether certain bushings could be made a little shorter. If a quarter of an inch [6.35 mm] could come off each one, they could be turned out considerably quicker. He knows this because he has been watching the clock [to gauge his productivity]. He knows it takes time to cut them, and he also remarks that the bushing is made of bronze and that bronze costs money. His suggestions therefore not only save time but considerable metal, since the bushings are turned out in large quantities. He is a specialist—a consulting engineer—in his little department, and his words will often mean money to you (The System Company 1911a, 16).

A kaizen blitz is an activity in which the workers find what Taylor called "the one best way" themselves instead having an efficiency expert supply it.[2] It's scientific management with worker participation. When the workers create the change themselves they internalize the underlying principles far more quickly. This is not a substitute for management commitment; the key idea is, "involve everybody as an active participant." Chapter 4 on lean manufacturing techniques will discuss kaizen blitz in more detail.

The Need for Management Commitment

The worst mistake that can be made is to refer to any part of the system as being "on trial." Once a given step is decided upon, all parties must be made to understand that it will go whether any one around the place likes it or not. In making changes in a system the things that are given a "fair trial" fail, while the things that "must go," go all right.

—Frederick Winslow Taylor (1911a, 136)

Do, or do not. There is no try.

—Yoda

The father of scientific management stated a key principle of organizational change, and George Lucas's fictional Jedi master rephrased it in an easily-remembered form. This principle applies to any quality or productivity improvement initiative. Commitment must come from top management. Bakerjian (1993, 1–13) cites management commitment as the most important element of any continuous improvement program. Feigenbaum (1991) warns that mere lip service or lukewarm support from top management is the "kiss of death" for any such program.

Do not expect instant results, although kaizen blitz is capable of delivering instant gratification on a local scale. Profound organizational change requires years. Some sources say that five years is not uncommon. Taylor (1911a, 195) cautions:

If any one expects large results in six months or a year in a very large works he is looking for the impossible. . . . But if he is patient enough to wait for two or three years, he can go among any set of workmen in the country and get results.

Performance Measurements Must Support Lean Manufacturing

Be careful what you wish for; you might get it.

—Old proverb, often illustrated in mythology and folk tales[3]

Lathin and Mitchell (2001) cite some outmoded paradigms and dysfunctional performance measurements that an organization must overcome to get a lean system. These include the idea that everyone and everything must keep busy by making parts. The chapter on the theory of constraints will show that 100 percent utilization should be a goal only at the constraint or bottleneck operation.

Standard and Davis (1999, 226–27) cite some examples of almost ludicrous stupidity—but they're apparently real. An engine plant in southeastern Michigan was measured against a goal of 1800 engines a day, even though the assembly plant required only 1600 a day. The plant manager dutifully built 1800 engines a day and when too much inventory piled up the plant closed for a week or so. A stamping company measured its purchasing department on getting the lowest prices. The transportation costs were, however, charged to the assembly plant that got the parts or materials. It was therefore possible to save a penny per unit on the purchased goods themselves while losing five cents per unit on transportation. Here is a real example from (apparently) the late 1990s:

> To keep idle machine tools busy and consume fixed overhead, produced parts were being manufactured regardless of customer demand. Result: New bottlenecks were created and WIP built up in front of them as the unneeded parts moved downstream. The inventory of unneeded parts grew, and the production of these parts competed for the same resources needed to produce customer-required parts. These unneeded parts consumed valuable machine time, scarce secondary processing resources, and assembly labor. This contributed to late deliveries (Voiland, 2001).

This manufacturer discovered that the best part about hitting oneself on the head with a hammer is that it feels great when one stops. A switch to the theory of constraints to guide decisions resulted in outstanding performance improvements.

Smith (1998, 17) cites a precious metals fabricator that measured three departments—the melt shop, rolling mill, and fabrication department—on the dollar value of their monthly shipments. Each area therefore grabbed the highest dollar-volume order available without regard to shipment due dates. Thirty-five percent of shipments were late. People often shipped nonconforming products even though they knew the customer would return them.

The incentive for each department to cherry-pick the best (under the performance measurement system) jobs resulted in inefficient resource utilization. Fifteen percent of the work took place during the first week of every month, and people often stood idle. Fifty percent took place during the last week and the company had to pay overtime. Be careful what you ask for, you might get it, and the manufacturing manager who imposed this measurement system did. It's doubtful, however, that what he asked (through the metric) and what he wanted (satisfied customers and efficient production) were the same.

Standard and Davis (1999, 228) define three standards for appropriate performance measurements:

1. Measurements must be objective, precisely defined, and quantifiable (measurable with numbers).

2. Performance must be under the control of the people or department that is being measured. This is also a requirement for what Joseph Juran calls "self-control."

3. The measurement must encourage behavior that helps the company meet its goals.

For example, maximizing equipment utilization may not even be under a department's control because lack of work from preceding operations can limit production. The theory of constraints shows that this will, in fact, often be true. The constraint limits the pace at which downstream operations can work. Measuring utilization also, of course, fails to meet the requirement that it help the company meet its goals. It promotes suboptimization (achieving high utilization by making unusable inventory) instead. The theory of constraints uses three measurements:

1. Throughput (finished goods *that are sold to customers*)

2. Inventory, which includes all investments in things the factory plans to sell

3. Operating expense

Cost accounting measurements such as labor variances and overhead do not appear anywhere among these. Standard and Davis (1999, 229–30) describe these measurements' effects in detail. Henry Ford warned long ago against letting "bean counters" run factories. Most forms of overhead are sunk costs; the factory incurs them no matter what it produces. Efforts to allocate overhead or absorb overhead by making as many units as possible are therefore futile at best and often dysfunctional.

Overcoming Self-Limiting Paradigms

> *Adherence to dogmas has destroyed more armies and lost more battles and lives than any other cause in war. No man of fixed opinions can make a good general . . .*
>
> —Major General J. F. C. Fuller (Tsouras 1992, 149).

One might wonder what the mythical hero Hercules has to do with lean manufacturing. The legends about Hercules were only stories but they may well have equipped Alexander the Great, who regarded Hercules as a role model, to conquer most of the known world—specifically by teaching Alexander to see through and overcome self-limiting paradigms. The transition to lean manufacturing also requires the organization to abandon paradigms, or preconceived ideas.

Many modern Hercules stories portray the hero as a Bronze Age version of Superman who uses his strength to rescue people, right wrongs, and so on. The original Hercules was impulsive and he sometimes killed bystanders in fits of rage (as did Alexander). His tasks often required him to find innovative solutions to seemingly impossible problems, for brute force could not have overcome some of his challenges. Table 3.1 shows the similarity between the thought processes of Hercules, Alexander the Great, and successful modern lean manufacturing practitioners.

Lean manufacturing allows us to similarly substitute brains for capital. We can get more capacity and shorter lead times without buying more equipment, although the latter is many people's first impulse.

Table 3.1   Hercules, Alexander the Great, and your factory: innovative thinking.

Situation	Self-Limiting Paradigm	Solution
Hercules' first Labor was to kill the Nemean Lion. When he attacked the lion, however, his arrows bounced off its hide.	A hunter must kill a lion by shooting it with arrows or stabbing it with a spear. Since the lion's skin was impenetrable, the lion could not be killed. Even if it could, it would be impossible to skin the lion.	Although the lion's skin was impenetrable, it had to be flexible (to allow the lion to move). Hercules slew it by strangling it. He then experimented and found that the lion's own claws would cut its invulnerable hide. He thus acquired a light but invulnerable garment. This story is, incidentally, why ancient Macedonian coins often depict Alexander the Great in a lion skin, and why Roman standard bearers often wore lion skins.
Hercules' next Labor was to slay the Lernean Hydra, which grew two heads for every one he severed. (Some factories' inventories seem to have this property.)	The Hydra was not only unkillable, attacking it made the problem worse.	Hercules told his nephew Iolaus to use a torch to burn each neck as soon as Hercules severed the head—an early form of *permanent corrective action*. Once the monster was dead, Hercules dipped his arrows in its deadly blood.

Note that Hercules acquired something (an impenetrable garment and poisoned arrows) from the first two problems he had to solve. The lesson is, "Look for a way to use everything."

continued

continued

Situation	Self-Limiting Paradigm	Solution
Hercules had to clean out the Augean Stables, which hadn't been cleaned for many years.	Hercules would have to do a lot of shoveling.	Hercules diverted two rivers through the stables, thus keeping his hands clean and finishing the job very quickly; a perfect example of working smarter, not harder.

The story of the Augean Stables teaches another lean manufacturing lesson: *Never take any aspect of the job, including its tools, materials, and methods, for granted.* Hercules was even ahead of Taylor, whose first inclination would have been to make the shoveling more efficient.

There was a legend that whoever unraveled the Gordian Knot would rule Asia Minor, but the knot could not be untied.	It was impossible to unravel the Gordian Knot.	Nowhere did the rules require the knot to be *untied*. Alexander drew his sword and cut it (Levinson, 1998. See also the movie *Alexander the Great*, with Richard Burton as Alexander).

The greatest obstacle is often mental. The person who knows something can't be done is always right—as far as he or she is concerned. Dr. Shigeo Shingo spoke of "nyet engineers" (nyet is Russian for "no"). Yoshiki Iwata referred to "concrete heads" (Womack and Jones 1996, 128–29), and the reference also cites "anchor draggers." The Carthaginian general Hannibal, on the other hand, said this of crossing the Alps: "We will find a way or make one."

Anyone who "knows something can't be done" had better hope the competitor knows it too. In 1940, France's generals thought German armor couldn't get through the Ardennes Forest (it did). The United States believed in 1941 that the water in Pearl Harbor wasn't deep enough for aircraft-delivered torpedoes (it was).

Alexander besieged the fortress of Tyre, which was on an island and therefore unapproachable by infantry.	Tyre was invincible.	Alexander may have remembered Hercules' alteration of geography (the Augean Stables) when he had his engineers build an isthmus from the mainland to the island, thus capturing the fortress.
Equipment setup is a non-value-adding activity.	Long production runs minimize setup times.	Single-minute exchange of die (SMED) does the same and it eliminates the need for long production runs.
We want to fill customer orders quickly.	We must keep a large inventory of finished goods.	Short cycle times allow the factory to make-to-order and reduce dependence on unreliable sales forecasts.

Elimination of Job Classifications

Getting an organization to "lean" requires elimination of restrictive job classifications, and cross-training to allow employees to do more than one job. Henry Ford encouraged every employee to regard anything that affected the organization's welfare as part of his or her job. Taiichi Ohno said that restrictive work rules, such as defining employees as lathe operators, welders, and so on, hinders the implementation of work cells and cellular manufacturing. In the Toyota production system a single operator might run a lathe, mill, and drill press. Cross-training is therefore essential to lean manufacturing.

Art Byrne of Wiremold adopted this approach (Womack and Jones 1996, 139–40). "He also knew that the existing work rules in Wiremold's union contract—restricting stampers to stamping, painters to painting, molders to molding, and so on—would make it impossible to introduce flow and to continuously improve every activity." Byrne overcame the union's objections by guaranteeing that improvements would not put anyone out of work.

Unions must accept the abolition of restrictive work rules, which really tie in with soldiering: a dysfunctional method of protecting jobs. Soldiering and work rules can protect jobs only until a competitor whose managers and workers know better put the company out of business. They cannot lead to sustainable higher wages. Schonberger (1986, 88) says that intelligent managers and union members understand that restrictive work rules are bad management. Bad management results in plant closures and job losses, so unions are more willing to accept relaxations in work rules.

"Lean" Does Not Mean Downsizing

> If any management attempts to effect savings in which it alone will benefit, it will and ought to fail. The reason that many plans involving personnel reductions fail is that the employer tries to reap the whole benefit (Basset 1919, 71).

Lean absolutely does not mean downsizing and making the remaining employees work harder for the same pay. Nor does it mean laying off half the workforce after improvements double the workers' productivity. These practices are demoralizing and destructive to worker commitment. They reinforce the dysfunctional behavior of soldiering, or deliberately limiting production, to protect jobs. Taylor recognized soldiering and he assigned most of the blame to management:

> . . . after a workman has had the price per piece of the work he is doing lowered two or three times as a result of his having worked harder and increased his output, he is likely entirely to lose sight of his employer's side of the case and become imbued with a grim determination to have no more cuts if soldiering can prevent it (Taylor 1911, 8).

Taiichi Ohno adds:

Hiring people when business is good and production is high just to lay them off or recruiting early retirees when recession hits are bad practices. . . . On the other hand, eliminating wasteful and meaningless jobs enhances the value of work for workers (Ohno 1988, 20).

Furthermore:

Slashing piece prices or reducing wages is a confession of weakness on the part of a foreman who indulges in these practices. Cutting out unnecessary operations or combining two or more which have been done singly is a proof of efficiency and can often be done with absolute gain to firm, workman, and buyer (The System Company 1911a, 25).

This is how the Ford Motor Company cut prices, increased wages, and earned more profit. Ford's labor relations went downhill quickly when its managers forgot or ignored its founder's principles. This is exactly what an employer must *not* do under lean manufacturing:

Twenty men who had been making a certain part would see a new machine brought in and set up [at the River Rouge plant], and one of them would be taught to operate it and do the work of the twenty. The other nineteen wouldn't be fired right away—there appeared to be a rule against that. The foreman would put them at other work, and presently he would start to "ride" them, and the men would know exactly what that meant (Sinclair 1937, 81).

This doubtlessly contributed to the United Auto workers' success in unionizing Ford despite its founder's insistence on a square deal for workers during the company's earlier days. Japan's National Railways had a similar experience in 1930, when the following labor leaflet appeared:

Brothers at the Omiya works!
How are your wretched lives these days?
It's intolerable that we can be driven 'til we drop by motion study, made to work for nothing by rate cutting, and threatened with being fired if we complain.
. . . Down with motion study! Down with rate cutting! (Tsutsui 1998, 53)

Employee turnover, whether through layoffs and early retirement incentives as described above, or worker dissatisfaction, undermines continuous improvement. Taylor (1911a, 62–63) says, ". . . a good organization with a poor plant will give better results than the best plant with a poor organization." He quoted a successful manufacturer as follows:

If I had to choose now between abandoning my present organization and burning down all of my plants which have cost me millions, I should choose the latter. My plants could be rebuilt in a short while with borrowed money, but I could hardly replace my organization in a generation.

Turnover (and incapacity or death) of a few key people broke the continuity of management principles at the Ford Motor Company in the mid-1940s. The result was that when Ford executives visited Japan in 1982 they were unfamiliar with the techniques they saw. "One Japanese executive referred repeatedly to 'the book.' When Ford executives asked about the book, he responded: 'It's Henry Ford's book of course—your company's book'" (Stuelpnagel 1993, 91).

Don't prove the Luddites right. Luddites were English textile workers who smashed automatic weaving machinery during the early 19th century because they were afraid they'd lose their jobs. If workers think productivity improvements will lead to layoffs they will do everything they can to make the improvements fail. Overt sabotage is not necessary, a "white mutiny"—doing exactly, and only, what one is told—is often quite sufficient.[4] If workers expect the improvements to lead to higher wages (Henry Ford's and Frederick Winslow Taylor's goal) they will not only cooperate, they will contribute proactively.

Womack and Jones (1996) stress this point repeatedly. *Do not lay off workers whom productivity improvements make unnecessary.* Reassign them, for example, to improvement teams or other work. The productivity improvements also reduce costs and allow the business to expand, thus creating more work. Wiremold's Byrne, for example, understood that "Resistance to continuous improvement would be chronic unless he guaranteed that workers would not be out on the street, even if their specific job was eliminated" (Womack and Jones 1996, 139).

This returns us to the importance of cross-training and the elimination of restrictive job classifications. Lean manufacturing often can and should get rid of certain jobs. The workers who were doing them must be willing and able to step into other jobs, which are often more demanding and require greater responsibility. Taylor (1911a, 146–47) says such progression is a natural result of scientific management (to which lean manufacturing adds greater emphasis on frontline worker participation).

Don't Worry about High Wages

Low labor costs are a dysfunctional incentive for manufacturers to move jobs offshore. Taylor and Ford both pointed out that high wages are not a problem when efficient manufacturing systems can afford to pay them. Taylor wanted "high-priced men," or workers whose skills and diligence justified high wages.

Basset (1919, 60) adds, "The efficient employer should pray for wages so high that his less efficient competitors will go out of business." Lean manufacturing is the way to get there. Basset reinforces another point that both Ford and Taylor stated:

> We all know that cheap labor is not cheap; paid cotton-pickers have proved cheaper than slaves—although it took a long time to convince the South, because they never reckoned the expense of idle slaves.[5] In any operation in which the material costs are high compared with the labor costs, the highest possible pay is the cheapest if it results in savings of material, or in a fine product, or in both. In the grades of production where labor is the big factor, high wages are economical if the wastes of human power can be kept to a minimum (Basset 1919, 64).

We have discussed the cultural transformation that must accompany physical techniques in an organization's transition to lean manufacturing methods. Lean enterprise comes from everyone, not only managers and engineers, thinking about ways to eliminate waste. Job security is a prerequisite for avoiding soldiering and malicious compliance; why should the workers help management eliminate jobs? Workers must, however, eschew rigid job classifications and embrace flexibility and cross-training.

"LEAN" APPLIES TO THE WHOLE ORGANIZATION

"Lean" applies not only to manufacturing but to all aspects of the organization, including transportation and delivery. Chapter 8 on supply chain management will show that to achieve the best results from lean manufacturing the entire supply chain must participate. A lean manufacturer cannot achieve its full potential in a chain of batch-and-queue suppliers and subcontractors. A lean shop floor cannot reap the full benefits of JIT if the purchasing department buys huge lots of materials to get so-called good bargains. This is why "lean enterprise" is better than just "lean manufacturing."

Get Rid of Bureaucracy and Red Tape

Bureaucrats and administrators cost money and they add no direct value to the company's product or service. Support personnel like managers, engineers, and technicians add value only indirectly by working through front-line workers.

Fairchild's Mountaintop plant places its self-directed work teams (SDWTs) in the inner circle of its organizational diagram (Wentz, in Levinson 1998, 70), whose central focus is the plant goals. The Society of Manufacturing Engineers' Computer and Automated Systems Association's New Manufacturing Enterprise Wheel places "people, teamwork, and organization" in its inner circle. This diagram's central focus is the customer.[6]

Tom Peters (1989) compared Wal-Mart's organization chart to its three-story headquarters. Sears' hierarchy was, meanwhile, more like the Sears Tower, which Sears actually had to sell. Nucor Corporation and Chaparral Steel have four layers of management. Peters (1987) notes that Lincoln Electric has one supervisor for every 100 employees. "Winners had 3.9 fewer layers of management than losers (7.2 versus 11.1) and 500 fewer staff specialists per $1 billion in sales." Henry Ford also urges readers to flatten the organization:

> To my mind there is no bent of mind more dangerous than that which is sometimes described as 'the genius for organization.' This usually results in the birth of a great big chart showing, after the fashion of a family tree, how authority ramifies. . . . If a straw boss wants to say something to the general superintendent, his message has to go through the sub-foreman, the foreman, the department head, and all the assistant superintendents, before, in the course of time, it reaches the general superintendent. Probably by that time what he wanted to talk about is already history (Ford 1922, 91).[7]

When Ford acquired the Detroit, Toledo, & Ironton Railroad during the early 1920s, he found a demoralized workforce and a business with a bad reputation. He abolished D.T. & I.'s legal department and cut legal costs by 93 percent. The railroad's entire executive staff occupied only two rooms, and a small building held the accounting department. D.T. & I. had 2700 employees before Ford bought it. Afterward, 2390 employees handled twice as much tonnage and they earned more money. After Ford had the railroad for a few years, its yearly earnings were about half what he paid for it, that is a 50% annual return on investment. This exemplifies Tom Peters' and W. Edward Deming's advice to flatten the organization.

ENDNOTES

1. Found on the Internet by a keyword search.
2. Kaizen blitz ("lightning continuous improvement") is a service mark per Cox and Blackstone (1998).
3. For example, William W. Jacobs' *The Monkey's Paw* (horror tale), and various folk tales.
4. "Sabotage" means literally "to clatter shoes," a French expression for working clumsily, i.e. soldiering.
5. This was written in 1919, when people were alive who remembered slavery.
6. http://www.sme.org/cgi-bin/new-gethtml.pl?/casa/casamwh.htm&&&CASA& as of 8/29/01
7. Carl von Clausewitz (1976, Book 5, ch. 5) says, "First, an order progressively loses in speed, vigor, and precision the longer the chain of command it has to travel. . . . Every additional link in the chain of command reduces the effect of an order in two ways: by the process of being transferred, and by the additional time needed to pass it on. It follows that the number of subdivisions with equal status should be as large as possible, and the chain of command as short as possible." Field Marshal Helmuth von Moltke (1800–1891) said essentially the same thing.

4

Lean Manufacturing Techniques

This chapter focuses on making jobs more efficient. It begins by describing the vital principle of friction, which the Japanese call *muda*, or waste. It then describes specific methods for improving tool and worker efficiencies.

Beware, however, of the word "efficiency." The theory of constraints will show that *the business must not measure individual operations on efficiency or utilization.* A goal of 100 percent efficiency or utilization is appropriate only for the constraint, or the operation with the lowest capacity. *Achievement of higher efficiencies at non-constraint operations does not increase the factory's capacity at all.* It does offer the following benefits:

1. Less labor time per piece, which frees workers for other activities

2. Shorter cycle times for individual jobs, which ties in closely with just-in-time delivery

   - The *lead time* is the time between an order's placement and its delivery. Harley-Davidson, which has extended its business to a wide variety of manufactured products, actually delivered and received payment for a product before it had to pay for the materials that went into it (Schonberger 1986, 191).

   - *Cycle time* is the time between a job's start to its completion. If the factory makes to order, lead time cannot be less than cycle time.

3. Cost reduction by elimination of unnecessary work such as straightening or fitting parts after heat treatment, or tapping bricks into place after setting them in mortar

4. Conversion from batch-and-queue operation to single-piece operation, for example by implementing single-minute exchange of die

Chapter 1 introduced the vital concept of friction, or inefficiencies that are not serious enough to demand immediate attention but the collective effect of which can degrade the enterprise's performance fatally. The next section provides more detail and it is perhaps the most important one in this chapter. Remember that a clear understanding of friction allows the definition of lean enterprise in a single word.

FRICTION: THE HIDDEN ENEMY

The Japanese call friction muda (waste) and muri (strain). The concept is so important that terms for it appear in many languages and contexts. *Lean manufacturing is about identifying and eliminating friction.* The first part, identification of the waste, is often as challenging as the second part. Even if the examples in this section do not relate directly to the practitioner's workplace, the thought process behind them will help him or her identify friction.

A criticism of Frederick Winslow Taylor was that he apparently took some jobs as givens. He almost quadrupled the amount of iron that workers could handle in a day and he raised their wages 60 percent. This was in itself a phenomenal achievement but *Principles of Scientific Management* mentions no investigation into whether conveyors or wheeled carts could move the iron. He apparently confined himself (via a self-limiting paradigm) to the assumption that the workers had to carry the pig iron.

Taylor also put a lot of work into improving the science of shoveling. He was well ahead of his time in prescribing different shovels for different materials (for example, iron ore versus ash or fine coal) when many people would use the same shovel for all jobs. The mythical hero Hercules, who used a shovel only to channel water through the Augean Stables, was ahead of Taylor. *Do not take the job, including its tools, materials, and methods, as a given.*

Something as simple as driving a screw is an example of unseen waste. It's especially insidious and subtle because value analysis would classify it as a value-adding activity. Cost of quality analysis would classify it as a required step. In manual assembly, however, "the cost of driving a screw can be 6–10 times the cost of the screw" (Cubberly and Bakerjian 1989, 5–9). The screwdriver must make a large number of turns and this takes a relatively large amount of time, even with a power tool. Don't take the screw as a given. Can a snap fit replace it? This kind of thinking drives waste from operations.

Gilbreth provides another example of waste that thrives right under our noses, or, more precisely, our writing hands.

One great aid toward cutting down the work of every one out of the trades as well as in, would be the standardizing of our written alphabet to conform to the laws of motion study. The most offhand analysis of our written alphabet shows that it is full of absolutely useless strokes, all of which require what are really wasted motions (Gilbreth 1911).

A practical problem with doing this, of course, is the difficulty in achieving a smooth transition between the existing alphabet and a new one. Shorthand actually addresses this problem. The need to transcribe speech drove its invention thousands of years ago:

In 4th century B.C.E. Greece, symbol systems were in use for abstracting alphabetic letters into single strokes that represented a letter, which, in turn, stood for common words, suffixes, and prefixes in which that letter appeared . . . (Encyclopedia Britannica 1909).

These inscription methods were used for business (recording deeds); education (recording lectures); law (recording judicial proceedings); religion (revising and annotating texts); and politics (recording orations) (C. Smith 1998).

The Romans adopted the Greek systems:

Cicero's freed slave, Tiro, is generally described as authoring the first truly tachygraphic ["swift writing"] system, one that more or less enabled verbatim reporting of human speech, by codifying a set of forms—symbols that by their shape and position represented common words. Tiro also developed a shorthand labor force, slaves trained to use the symbols who were positioned around the forum in Rome to record debate in relays (C. Smith 1998).

The familiar QWERTY keyboard is another example of (possible) waste that may thrive right under our fingers. It was developed in the 19th century to prevent jamming of then-mechanical type bars. It achieved this by separating commonly-used letters. (Internet references suggest that the story that it prevented jamming by deliberately slowing the typists may be a myth.) Today, of course, there are no mechanical type bars to jam. A more efficient arrangement may be possible but, like the alphabet, QWERTY is an ingrained standard. We do not contend that it is practical to replace

QWERTY with something else, but this is another example of how waste can settle into a job and appear perfectly natural. The example's purpose is to reinforce the *way of thinking* that leads to recognition of waste.

Types of Waste

Bakerjian (1993, 9-2) and Ohno (1988, 19–20) cite seven types of waste:

1. *Overproduction.* This is often a consequence of push-style production control and of performance measurements like equipment utilization and cost-per-piece. Do not let the cost accounting system run the factory.

2. *Waiting, time in queue.* Variation in processing times leads to a hurry-up-and-wait effect that causes inventory to accumulate even in systems that supposedly have extra capacity. Large production batches also lead to waiting time.

3. *Transportation.*

4. *Non-value-adding manufacturing processes.* Example: Ford had a process to straighten axles after heat-treatment. The employees identified this correctly as not a value-adding operation but 100 percent rework. They found a way to cool the axles without deformation, thus eliminating this procedure.

5. *Inventory.* It sounds obvious but it isn't. Companies have instituted just-in-time warehouses (an oxymoron) to supply customers who want JIT delivery. A freight ship that carries containers is a very cleverly disguised warehouse. *Float,* or materials in transit to factories, can be very substantial inventory.

6. *Motion.* Elimination of waste motion was a principal goal of the early scientific management practitioners.

7. *Costs of quality: scrap, rework, and inspection.* Suppression of this form of waste also supports ISO 9000 and quality assurance.

False Economy Is Waste

Avoiding false economy, or "cheap is dear," is another waste reduction concept. Excessive emphasis on cheap labor, cheap materials, and cheap equipment can lead to ruin. Taylor tried to seek out "high-priced men," or employees whose willingness and ability to follow instructions made them worth high wages. Henry Ford (1930, 53) said that good workmanship was cheap even if one had to pay more for it because first-class workers would get the most use out of the capital equipment. Looking for cheap workers could ruin a business as quickly as buying the cheapest materials.

The machining trade of the late 19th century provides another example of false economy. Frederick Winslow Taylor and Maunsel White, a metallurgist at Bethlehem Steel, showed that tools made from new alloy steels could cut metal several hundred percent faster than was previously possible:

> . . . they had upset one of the most hallowed precepts of the machinist's craft: the belief that the proper cutting speed was the one that maximized the life of the cutting edge of the tool. Not so, said Taylor: tools can and should be reground regularly and systematically; what we should maximize is the amount of metal removed per unit of time, and cutting speeds should be set accordingly (Aitken 1960, 30).

Efforts to make the tools last longer avoided one form of waste but created a worse one: lower productivity per unit time. In this case the waste was in the method but the same principle applies to personnel, equipment, and materials. Henry Ford hired only the best sailors for his cargo ships and he paid correspondingly high wages. The true priority was to get the most use out of the ship, which represented a huge capital investment. Under this consideration a first-class crew wasn't expensive, it was cheap.

Modern factory machines have electric motors but the belt-driven machinery used in the past offers us another example of false economy. The System Company (1911a, 88–89) notes that leather belting could be purchased for 10 to 12 percent below the cost of first-class goods. The leather that went into these cheaper belts, however, came from too far from the center of the animal hide. This resulted in thin and soft spots that, if they were on the belt's edge, caused it to stretch unevenly. Such belting would not run true on pulleys. It was unlikely to last three months in applications where first-class belting might last a year. The authors conclude, "It is the poorest of economy to save ten percent by putting in belts below the standard in quality."

We will later discuss in detail how Omark, a saw chain manufacturer, eliminated heat treatment by buying an alloy that did not need it. Whether the metal was more expensive is not known but it eliminated a time-consuming process step.

The Frontline Worker's Role in Reducing Friction

Halpin (1966, 60–61) discusses the error cause removal (ECR) system, which was developed at General Electric, for quality improvement. Examples of error causes include:

- A poorly placed lamp.

- A left-handed operator having to use a machine that was set up for right-handed workers. (This is an ergonomics or human factors issue.

- A telephone on the wrong side of a desk.

Halpin summarizes the key point: "They turned out to be the *little things that get under a worker's skin but are never quite important enough to make him come to management for a change"* (author's emphasis). Levinson and Tumbelty (1997, 24) also provide a one-sentence definition to help frontline workers identify friction. *"If it's frustrating, a chronic annoyance, or a chronic inefficiency, it's friction."*

Edward Mott Woolley made the following observations about friction (although he did not use this term) in stenography. These are more examples of chronic annoyances that do not prevent performance of the job, but make it harder for the worker and less cost-efficient for the employer:

Many of the typewriters' tables were several inches too high; chairs were not suitable; lights were insufficient or badly placed; no scheme of arranging desks had been followed to reduce steps; no system was in effect for the care and maintenance of machinery. Furthermore, no standard of machine had been adopted: half a dozen makes were in use (The System Company 1911, 44).

Schonberger (1986, 19–21) suggests that manufacturing workers record the reason for *any* work slowdown or stoppage. This is the equivalent of a yellow light in a yellow–red andon (light panel) system, with red meaning an actual line stoppage. The problem and its cause should be recorded immediately. Causes can then be summarized on check sheets or tally sheets and prioritized on Pareto charts, thus using two traditional and easy-to-use quality improvement tools.

Related tools include ECR (Halpin 1966) and the quality/productivity *hiyari*, or scare report (Imai 1997). Both are worker-initiated activities to remove a source of friction (or make some other correction or improvement). This chapter devotes an entire section to the Ford Motor Company's team-oriented problem solving, eight disciplines (TOPS-8D) technique, a systematic process for identifying a problem's root cause and implementing a permanent correction. It is extremely versatile and it should be adaptable to *hiyari* or even improvement suggestions.

This section has covered friction (muda or waste, muri or strain) in job design, and it has provided a one-sentence definition for frontline workers to use. The next section considers efficient and inefficient use of materials, which relates closely to friction because inefficient use of material is also easy to overlook.

EFFICIENT USE OF MATERIALS

He perfected new processes—the very smoke which had once poured from his chimneys was now made into automobile parts.

—Upton Sinclair, *The Flivver King* (1937, 61)

The concept of using materials efficiently—really efficiently—ties in with that of reducing friction. Another of Henry Ford's success secrets was the ability to get the most out of materials. There is even a legend that he asked a supplier to package its products in wooden boxes with boards of specific dimensions—those of Model T floorboards. The vendor would have had to package the product anyway and probably didn't care what the board dimensions were.

The difference was, however, that the box went into a product instead of a wood distillation plant. This was, incidentally, how Ford preferred to deal with wood that could be put to no other purpose. He burned combustible wastes for heat and power but only as a last resort. He insisted on recovering the valuable chemicals from coal before burning the remainder (coke). Coal cost five dollars a ton upon delivery to the factory but, after conversion into coke and byproducts, it was worth twelve dollars a ton (Ford 1926, 105–6). The process, incidentally, removed much of the sulfur from the coal (saleable as ammonium sulfate, a fertilizer) long before people recognized acid rain (the result of burning high-sulfur coal) as a problem.

Ford grew up on a farm and he may have remembered the meatpackers' admonition, "Use everything but the squeal" (Figure 4.1). This is not a comforting thought for anyone who likes sausage but it conveys the general idea: find a use for everything.

Australian farmers are extending the "use everything but the squeal" concept by vaccinating animals against the methane-producing bacteria

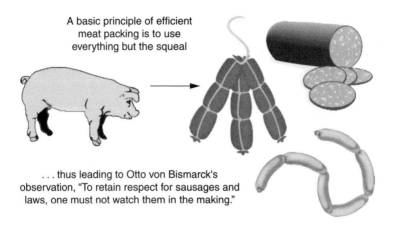

A basic principle of efficient meat packing is to use everything but the squeal

... thus leading to Otto von Bismarck's observation, "To retain respect for sausages and laws, one must not watch them in the making."

Figure 4.1 How to think about materials.

that live in the animals' digestive systems. The idea is to have the animals, not the bacteria, use the food. This is another example of waste reduction.

Levinson (2002) shows that Henry Ford's strategy of wasting nothing and seeking a use for everything would have allowed his businesses to meet ISO 14000 requirements long before anyone had ever heard of ISO 14000 (or the Environmental Protection Agency). This was during an era when he could have legally thrown into the river anything that wouldn't go up the smokestack. As shown by the excerpt from *The Flivver King*, he realized that *money* was going up the smokestack and he found ways to recover it. Lean manufacturing and lean thinking therefore complement and support ISO 14000.

The idea of recognizing waste wherever it appears cannot be overemphasized. The following example is extremely useful because it shows how Henry Ford could apparently recognize waste on sight. We want everyone in the organization to think the same way:

> One day when Mr. Ford and I were together he spotted some rust in the slag that ballasted the right of way of the D. T. & I. [railroad]. This slag had been dumped there from our own furnaces.
>
> "You know," Mr. Ford said to me, "there's iron in that slag. You make the crane crews who put it out there sort it over, and take it back to the plant" (Bennett 1951, 32–33).

Most people would probably have overlooked this because rust belongs with the other junk. Ford realized that (1) rust contains iron and (2) it was in the slag from his blast furnaces. Taking it back to the plant was only containment, the third step of the 8D problem-solving process that this chapter will discuss later. Permanent correction meant keeping the iron from getting out of the factory in the first place. The steel mill installed powerful electromagnets for removing particles of iron from blast furnace slag.

Here is another example of the need to *see the waste*. As with rust, most people wouldn't give smoke from a smokestack a second thought. Smoke is black and one expects it to come from smokestacks, so what's the problem?

Black smoke is unconsumed carbon—nascent heat—lost energy—wasted coal. A smoking chimney registers money lost (The System Company 1911a, 28).

Today, black smoke might suggest noncompliance with environmental protection regulations. The obvious way to deal with it is to install cyclones, baghouses, precipitators, or other particle removal equipment, along with scrubbers for toxic gases. This results in capital outlays that yield no return. The less obvious but far superior approach is to avoid making the black smoke in the first place, for example by assuring complete combustion and thus getting more heat per unit of fuel. Ford extracted the valuable chemicals from the coal before he burned the remaining coke. Sulfur that would today require scrubbing (to remove sulfur dioxide, a toxic gas that also contributes to acid rain) became fertilizer. This is the kind of thinking that everyone should use in any enterprise. *Remember, if the waste was obvious, someone would have already done something about it.*

Design Parts and Processes for Waste Reduction: Make Parts, Not Chips and Shavings

Mege (2000) discusses high agility machines (HAMs) that shape aluminum parts. An American purchaser of a HAM boasted that it ate 400 tons (364 metric tons) of aluminum every month. This customer had to buy a second aluminum chip compactor, and it sells bricks of compacted aluminum chips to recyclers. Furthermore, the finished part's weight is often only 15 to 20 percent of that of the original metal billet.

The idea is to make parts, not bricks of aluminum chips. We don't know enough about the actual product to say if there is a way to make it more efficiently. One might ask, however, "Is there another way to make these parts so they won't need as much machining? Why do we have to grind the equivalent of five out of every six billets into chips, which are then sent to a recycler to be made back into billets?" This story shows again that waste is rarely obvious. Machining is a value-adding activity and, under cost of quality analysis, a required activity.

Redesign of parts or manufacturing processes can reduce the need for machining. Although machining is often the best way to achieve tight tolerances or smooth finishes, it is also inherently wasteful.

The degradation of materials by conventional machining methods is of the order of 30 to 70 percent, and the more complex shapes are at the higher end. Most of the chips are recycled as scrap, but there is a severe economic penalty. As a result, there has been increasing emphasis on 'chipless machining' processes by which a part is made to final, or near-net, shape. Precision forging, precision investment casting, and powder-processing techniques are good examples of such processes (Dieter 1983, 210–11).

Jacques (2001) describes the manufacture of wooden baseball bats at Hillerich & Bradsby Company (H&B), the maker of the Louisville Slugger. Two pictures draw immediate attention. In the first, someone is holding a cylindrical wooden billet. Lean thinking includes visualizing the bat inside the billet. The next picture confirms the reader's expectations: a lathe with a lot of sawdust and shavings. The billets weigh 80 to 100 ounces (2268 to 2835 grams) when they arrive in Louisville. The bats weigh 30 to 35 ounces (850 to 992 grams). Fifty-six to 70 percent of the material is therefore wasted.

The idea of cutting two or more bats out of a slightly larger billet (which may or may not be cylindrical) as shown in Figure 4.2 therefore comes to mind. This is probably harder than cutting parts from a flat board, although the 19th-century Blanchard lathe could form complex wooden objects like musket stocks. The roughly-shaped bats would then require turning on a lathe to get the desired surface finish.

This does not necessarily mean there is a better way to do this particular job. Product quality, equipment capital costs, or both may take precedence

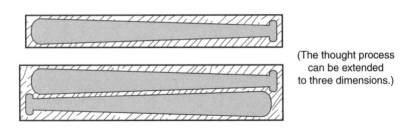

(The thought process can be extended to three dimensions.)

Figure 4.2   How many bats are there in a wooden billet?

over the cost of the material. The material cost is almost certainly negligible for a professional bat (H&B custom-manufactures them for pros) but perhaps not for retail ones.

Some very helpful information was provided by Mr. Jack Hillerich at H&B. It turns out that the company once *did* nest baseball bats to get more out of each piece of wood. This improved material usage by about 20 percent. Today, however, the company uses tracer lathes that use single knives to cut entire bats. This requires the lathe to hold the bat in a steady ring that precedes the knife. A piece in the rough shape of a bat will not work in this fixture.

A roughly-shaped bat could be machined by a more complex lathe with as many as 40 knives. Setup times are long and this results in long production runs. This works against the lean manufacturing principle of short setups and short production runs. The tradeoffs that would be necessary to use more of the wood exceed the cost of the material.

Similar considerations might apply in the HAM example. Aluminum is a relatively inexpensive metal. It might not be worth the addition of extra equipment or processing steps to get more out of each aluminum billet although the economics could change easily for a costly specialty metal or alloy. *The key point is that any pile of sawdust, metal turnings, or chips should attract attention in any machining operation.* It's possible that there's no way to avoid making that pile but it at least deserves a look.

Implications for ISO 14000 and Greenhouse Gas Reduction

Recall that lean manufacturing is synergistic with ISO 9000 and ISO 14000. Compliance with the ISO 14000 environmental standard should be not only free but profitable. There are two ways to approach ISO 14000: throw money at it or make money from it. The money-throwing approach might include paying for more scrubbers or waste treatment plants to reduce pollutant streams. The smart approach is to not make the waste in the first place or, as Ford did, find ways to turn it into marketable products.

Global warming due to greenhouse gases like carbon dioxide and methane may or may not be a real issue.[1] Either way, ratification of the Kyoto Protocol would be an economic catastrophe for the United States. It would raise energy costs for both manufacturing and transportation, thus making American goods less competitive with those from developing nations that have no obligations under the treaty. California's experiences in 2000–2001 show what Kyoto would do to the entire country.

As an example, energy-intensive aluminum plants on the West Coast had long-term contracts for fixed-price electricity. They found it more profitable to shut down and resell the electricity to California. Dysfunctional economic driving forces therefore turned value-adding manufacturing plants into non-value-adding middlemen, a trend that is very harmful to our nation's welfare. The blame lies not with the aluminum factories but with California's policy of discouraging new power plant construction.

Furthermore, many employers are considering leaving California because rolling blackouts keep interrupting production. Other states would be delighted to get the manufacturing jobs and they are soliciting California employers, many of whom are about ready to go:

> At the epicenter of the energy crisis, STC Plastics (Chino, California) has so far experienced more than 20 outages this year, says president Scott Keller. . . . If the trend continues, he may consider moving his operations to another state with lower power costs.
>
> . . . "We got into a mess that has nothing to do with regulation, and everything to do with politics," says Earl Bouse, vice president of manufacturing services of Hanson Permanente Cement (Pleasanton, California). . . . "If state agencies like Caltrans won't pay more for California-made cement because electricity costs have gone up, then will they be tempted to buy cement from countries like Korea, China, or Thailand where electricity is cheaper?" (Kim 2001)

Bouse adds that it takes about 130 kWh of electricity to make a ton of portland cement. The same article shows that the problem has spread beyond California because other West Coast states are on the same power grid: Georgia-Pacific Corporation permanently closed its pulp mill and chemical plant in Bellingham, Washington.

Chlorine and sodium hydroxide are manufactured from salt water by electrolysis. The Niagara Falls, New York, area was a popular location for siting chlor-alkali plants for the same reason that bauxite (aluminum ore) may be shipped from Australia to Scandinavia for processing: relatively cheap hydroelectric power. Atofina Chemicals, Inc. suspended operations at its Portland, Oregon chlor-alkali and sodium chlorate plant because of high electricity costs. An online news release at the company's Web site, www.AtofinaChemicals.com, cites "unprecedented power prices in the Pacific Northwest"—prices driven both by California's energy policies and low rainfall that reduced hydroelectric power capacity.

The economic driving forces of electrical power costs should now be quite obvious. Henry Ford said that *power and transportation costs* (not, incidentally, access to cheap labor) *are the two paramount considerations in siting a new factory.* Ratification of Kyoto would encourage employers to move the smokestacks, all their carbon dioxide, and the manufacturing jobs under the smokestacks offshore to places that have no obligations under Kyoto—thus ruining the American economy while doing nothing for a problem that may or may not exist.

It would, in fact, worsen the global environment. U.S. environmental regulations, like those in most other industrialized nations, limit the release of materials that are harmful to human, animal, and plant life. Although the U.S. will hopefully not regulate carbon dioxide emissions, it is looking very closely at mercury emissions from coal-fired power plants. The U.S. already limits sulfur dioxide emissions because this gas is toxic and it causes acid rain. Most developing nations have few if any environmental regulations, so moving the smokestacks offshore would add poisons and acid rain to the greenhouse gases.

The Kyoto Protocol includes emission trading, another dysfunctional economic driving force. There is an entertaining story about an entrepreneur who learns that the United States Government pays farmers to not raise hogs (to avoid overproduction and low prices for farmers). He writes to the Secretary of Agriculture, states his intention of going into the business of not raising hogs, and asks what kind of hogs are best not to raise. He also asks what kind of farm is best for not raising hogs, and so on. Then he adds that the hogs he doesn't raise won't eat any corn, and asks whether he can also get paid for the corn that need not be grown. This story reduces to absurdity the idea that people should be paid for not producing wealth. National wealth and prosperity come not from trading carbon dioxide credits or "not raising hogs," but from making and selling products.

There are, however, economic driving forces that have the *incidental* effect of reducing carbon dioxide emissions. Green manufacturing and greenbacks are quite compatible under intelligent environmentalism, as shown in Table 4.1. The intelligent methods use fuel cells, which convert chemical (fuel) energy directly to electrical energy. This bypasses the efficiency limits of power generation cycles that transfer energy from hot reservoirs (furnaces, via boilers) to cold ones (condensers). The bottom line is that one can get about twice as much useful energy per unit of fuel. The user pays less for fuel (the economic driving force) and also happens to generate less carbon dioxide. The first companies to implement practical fuel cell technology are going to make a lot of money.

This section has shown how lean thinking—the idea of getting the most from every resource—applies to areas as diverse as ISO 14000, environmental protection, fuel conservation, and greenhouse gas reduction. Lean thinking enables its practitioners to get more while paying less, as opposed to using throw-money-at-the-problem approaches that require little creative (or other) thought. The next section introduces *standardization*, the implementation of uniform work instructions and specifications that make productivity and quality improvements permanent. Standardization supports what Joseph Juran calls, "holding the gains."

Table 4.1   Smart and not-so-smart environmentalism.

Pay More and Get Less	Green Plus Greenbacks
The zero-emission vehicle: Electric vehicles with batteries that require frequent (and therefore inconvenient) recharging. It doesn't get rid of the exhaust pipe either, it moves it from the vehicle to the power plant.	Systems are under development that can crack hydrogen from conventional fuels like gasoline and diesel fuel. The vehicle's fuel cell then converts the hydrogen, with efficiencies of about twice those of the Otto (internal combustion) power cycle, into electrical power.
The zero-emission vehicle: Cars and buses that burn (or even use in fuel cells) hydrogen. The hydrogen must be under enormous pressure if the fuel tank is to hold enough for practical purposes. The *Hindenburg* airship also used hydrogen (for buoyancy), and we know what happened to the *Hindenburg*. It might be possible, however, to adsorb hydrogen in solids for safe storage and transportation. A search for "hydrogen storage" and "adsorption" on the Internet points to experiments with carbon nanotubes.	This will allow construction of vehicles that get almost twice as many miles or kilometers per gallon or liter of fuel with no reduction in weight or power, and the fuel of which is not a compressed flammable gas. (Very little hydrogen is actually present in the system at any time because it's created only as it's needed.)
	Aerodynamic streamlining of vehicles increases fuel economy without trading off safety or performance. Overdrive transmissions reduce fuel consumption on relatively flat highways.
Higher corporate average fuel economy (CAFE) standards that would compel manufacturers to make smaller, less safe, and less desirable vehicles.	Don't do this: $C + O_2 \rightarrow CO_2 +$ heat In practice, a steam turbine can convert perhaps 35 or 40 percent of the energy in coal into useful work.[2]
Install expensive scrubbers to remove carbon dioxide from smokestacks, and pay extra money to sequester the carbon dioxide.	Do this instead: $C + H_2O \rightarrow CO + H_2$ and also $CO + H_2O \rightarrow CO_2 + H_2$
Coal-burning power plants already use scrubbers and cyclones to remove *incidental* combustion products like particles, sulfur and nitrogen, and oxides, all of which endanger human health, wildlife, and crops. Carbon dioxide is the *principal* combustion product and its removal would increase power costs enormously. On the other hand, it is not harmful to plants or animals in the concentrations that come from smokestacks.	Fuel cell $2H_2 + O_2 \rightarrow 2H_2O +$ electricity The technology for making hydrogen from coal (carbon) and steam has been around for a long time. It's thermodynamically possible to get about twice as much electricity per unit of coal this way because fuel cells are not subject to the efficiency limits of thermal power cycles.
Tax methane-producing sheep and cattle because "Over half of New Zealand's greenhouse gases comes from the country's 46 million sheep and nine million dairy cows and beef cattle."[3] Methane, the product of bacterial digestive processes inside sheep and cattle, is 21 times as potent a greenhouse gas as carbon dioxide.	Australia is trying a drug or vaccine that suppresses methane-producing bacteria. This is very sensible because the idea is to turn animal feed into beef, mutton, or wool, not methane and bacteria ("Australia fights methane" 2001). It yields more animal products per unit of food and it has the incidental effect of reducing greenhouse gas emissions.

STANDARDIZATION

The stereotype that scientific management required workers to leave their brains at the factory gate comes largely from Taylor's emphasis on standardization. Standardization means the implementation of the best way of doing a job. Best practice deployment means implementing this standard throughout the organization.

Standardization went against the craft and trade tradition that the worker should do the job according to his or her judgment. Scientific management looked for the best way to do the job and then standardized this method throughout the establishment.

Standardization led to knowledge retention. In the past, craft and trade workers may have improved their methods but they didn't share these improvements (best practice deployment). They often didn't record them either so when workers died or retired the knowledge went with them. The same principle appears in modern references:

> . . . action must be taken to ensure that improvements are incorporated into the daily routine and become a permanent part of the standardized procedure. Otherwise, the improvements will erode quickly and the benefits will be lost. Improvement without standardization cannot be sustained (Standard and Davis 1999, 117).

Juran and Gryna (1988, 22.57–22.58) describe the *knack*: "a small difference in method which accounts for a large difference in results." A knack may result in superior performance by the workers who use it. They sometimes conceal the knack to avoid criticism for not following the standard procedure, or to retain an edge over other workers. The latter consideration is an argument against ranking workers against each other. Negative knacks, on the other hand, are unwitting deficiencies in technique that reduce productivity or quality. Retention and dissemination of beneficial knacks and elimination of negative knacks are reasons for standardization. Juran and Gryna (1988) provide two examples:

1. An investigation of crankshaft damage showed that only one worker's product was involved. He sometimes bumped the crankshaft into a nearby conveyor because he was left-handed, and the workplace was designed for right-handed workers. (This is not exactly a negative knack, but rather a human factors and workplace design issue. This subject will be discussed in more detail later.)

2. Only one worker in an aircraft assembly plant met his production quota consistently. He had taken his powered screwdriver home and rebuilt its motor. When the company did the same with the other screwdrivers, productivity went up.

Juran and Gryna advocate incorporation of beneficial knacks into the process and error-proofing to preclude negative knacks. This is simply standardization and best practice deployment. Today, of course, the ISO 9000–required documentation system provides knowledge retention. Standardization is therefore a simple example of synergy between lean manufacturing and ISO 9000. The work instruction or operating instruction is standardization; workplaces that are in the forefront of worker empowerment use this aspect of Taylorism today.

Standardization reduces variation because everyone does the job the same way. It also supports continuous improvement (kaizen) by taking the backward step out of the two-steps-forward-and-one-back effect. This is exactly the point that Taylor raised.

Taylor foresaw the concept of the modern business school by applying the knowledge retention principle to professional or "white-collar" workers:

> Unfortunately there is no school of management. There is no single establishment where a relatively large part of the details of management can be seen, which represent the best of their kinds. The finest developments are for the most part isolated, and in many cases almost buried with the mass of rubbish which surrounds them (Taylor 1911a, 200–201).

Taylor recognized and supported the idea that frontline workers could suggest improvements. If such a suggestion worked, it, like any other improvement, became the new standard or "one best way." Scientific management did, however, separate planning from production, and it looked to the planners for most improvements. Modern lean manufacturing recognizes a far greater role for the frontline workers or doers.

Best Practice Deployment

Harry and Schroeder (2000) stress best practice deployment as a key element of Six Sigma. If an improvement can apply to different operations it should be deployed across the company as a "best practice." The concept actually dates back to Taylor:

> It is true that with scientific management the workman is not allowed to use whatever implements and methods he sees fit in the daily practice of his work. Every encouragement, however, should be given him to suggest improvements, both in methods and in implements. And whenever a workman proposes an improvement, it should be the policy of the management to make a careful analysis of the new method, and if necessary conduct a series of experiments

to determine accurately the relative merit of the new suggestion and of the old standard. *And whenever the new method is found to be markedly superior to the old, it should be adopted as the standard for the whole establishment.* The workman should be given the full credit for the improvement, and should be paid cash premium as a reward for his ingenuity (Taylor 1911, 67, emphasis is ours).

Hewlett-Packard's "best practices" program almost quotes Taylor:

Each job at HP has standard operating procedures (SOPs) associated with it. These SOPs are called practices. If an employee can find a better way to accomplish a job, he or she is encouraged to document the new method. Management then reviews the new method and its estimated advantages and, if approved, it becomes a new "best practice." Each procedure guide is updated and the employee receives credit for the new method. Appropriately, the reward system is geared to benefit those employees who actively seek a better way to do their own jobs (Bakerjian 1993, 1–8).

In summary, standardization and best practice deployment prevent the "two steps forward and one back" experience that is common in many businesses. Standardization holds the gains and best practice deployment spreads the improvement's benefits throughout the organization.

The greatest leverage for achieving leanness, however, is often in the design stage and not on the shop floor. A good design confers many cost-reduction opportunities on the shop but the best manufacturing teams in the world might not be able to overcome the waste inherent in a bad design.

PRODUCT DESIGN PRINCIPLES

Lean manufacturing begins with the product design. Design for manufacturing (DFM) and design for assembly (DFA) are the key concepts, and they also play a role in the design aspect of ISO 9000. Lean manufacturing and ISO 9000 are again mutually supporting and synergistic. Cubberly and Bakerjian (1989) describe the following principles for DFM and DFA.

1. *The design should have as few parts as possible. We want the opposite of a "Rube Goldberg" design.* "Start with an article that suits and then study to find some way of eliminating the entirely useless parts. This applies to everything—a shoe, a dress, a house, a piece of machinery, a railroad, a steamship, an airplane. As we cut out useless parts and simplify necessary ones, we also cut down the cost of making" (Ford 1922, 14).

2. *Use modular designs.* Schonberger (1986, 154) says that designing modular products with low part counts makes it easy for the shop to solve any subsequent problems. Complicated processes and products have the opposite effect.

3. *Use standard parts and components.* "Even more important, standard parts are proven parts. In the past, Xerox typically put 80 percent newly designed components into a new model of copier. The results: a long design cycle followed by a long debugging cycle and tardy entry into the marketplace. Xerox's new 9900 copier used only 30 to 40 percent new components, which helped cut the design–to–market time in half" (Schonberger 1986, 154).

4. *Use multifunctional parts and include features that help with assembly.* Such features include keys that permit assembly only in the correct orientation. This is a form of poka-yoke or error-proofing. Other features might include reflective surfaces that make inspection easier.

5. *Fasteners (bolts, nuts, rivets, and screws) are not your friends, nor are they the shop floor's.* The cost of driving a screw can be five or ten times as much as the cost of the screw. Can snap fits replace fasteners?

   • When fasteners are necessary there are techniques for making assembly, and disassembly for maintenance, far more rapid. If you are unhappy with a screw that seems to turn forever before it tightens or comes out, your customer (or the shop floor worker) is likely to be unhappy too. A clamp that loosens and tightens with a couple of turns of a bolt is more convenient than a bolt that must be driven completely through a flange or other closure. Robinson (1990, 337) describes the split-thread bolt, which tightens with a sixty-degree turn of the wrench.

   • Henry Ford was enthusiastic about welding forged and stamped parts, instead of riveting them together.

6. *Assemble everything from one direction if possible.* Workers and equipment should not have to reposition themselves or the work (non-value-adding activities). There is nothing wrong with designing a job so people can work on a piece from two sides at once. It is quite practical to have a machine work a piece from four sides. Arnold and Faurote (1915, 80) show a Foote-Burt machine that drills 45 holes at once in a cylinder casting from four directions. They also show another Foote-Burt machine that

taps 24 holes simultaneously from three directions (p. 82). The key idea is that there is no repositioning of the part itself.

7. *Compliant parts do not require excessive force for assembly.* Tapers, chamfers, leads, slots, and flats help guide parts into place.

8. *Minimize handling and adjustment.* "Handle," "orient," and "adjust" are all non-value-adding activities. Handling also creates opportunities for defects.

9. *Avoid flexible components like cables and wires.*

   • If they are necessary, color-code them (and remember that some workers are color-blind) or include other features that prevent connection to the wrong place. Slots and keys prevent backward insertion; the polarized plug is a very simple example.

   • Use circuit boards if possible.

We have already seen that it also is helpful to design parts or processes to avoid extensive machining.

Value Analysis and Quality Function Deployment

Juran (1992, 192–93) defines value analysis as "a process for evaluating the interrelationships among (a) the functions performed by product features, and (b) the associated costs." The idea is to design the product to deliver its essential functions at the lowest cost. Elimination of marginal product features can hold costs down.

Value analysis inputs include:

• Customer requirements, and their importance

• Corresponding product features that satisfy the customer requirements

• The product features' estimated costs

• Information on competitive products, their features, and their costs

Notice the tie-in with quality function deployment (QFD), also known as the "house of quality."

Figure 4.3 shows the major parts of a QFD diagram (described by Hradesky 1995, 663). The diagram can also include column (product characteristic) sections for technical difficulty, benchmarking, and target values or requirements. The rows (customer requirements) can be extended to include competitive evaluation.

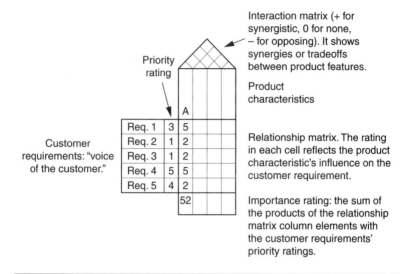

Figure 4.3 Quality function deployment.

The figure shows a sample calculation for product characteristic A. The interaction matrix shows tradeoffs between product characteristics. For example, selection of titanium instead of steel would have a positive interaction (synergy) with lightness and a negative interaction (tradeoff) with ease of machining. Titanium chips can ignite in air so machining of titanium requires special precautions.[4]

The importance ratings for customer requirements (including safety and conformance to government regulations) also tie in to a failure mode and effects analysis (FMEA) through failure mode severity ratings. For example, a failure mode that endangered the user would have a very high severity rating. A failure mode that involved a low-priority customer requirement and therefore would cause only minor annoyance would have a low severity rating. The connection between value analysis, QFD, and FMEA reinforces the point that quality and productivity improvement methods are synergistic and mutually supporting, not stand-alone, activities.

A value analysis spreadsheet (Juran 1992, 194) is a matrix, the rows of which are the product features and their costs (Table 4.2). The columns are their functions—best described by a verb and a noun. The matrix cells contain the allocation of each feature's cost to its function. Juran warns that this is not an exact exercise, and estimates are often made to the nearest 10 percent.

This section has covered the role of design for manufacture (DFM) in reducing manufacturing costs. A manufacturing-unfriendly design can preclude a lean process no matter how good the factory is. Value analysis

Table 4.2   Value analysis spreadsheet.

Product Feature	Cost (and % of Total)	Functions (verb and noun)		
		Ease Maintenance	Promote Reliability	Improve Safety
Specialty material	$20.00/40%		$15.00	$ 5.00
Part fabrication	$20.00/40%	$ 5.00	$ 7.50	$ 7.50
Assembly	$10.00/20%	$ 8.00	$ 1.00	$ 1.00
Total cost	$50.00	$13.00	$23.50	$13.50
Percent		26%	47%	27%

and quality function deployment (QFD) are tools for integrating the customer's needs with the design. On a final note, DFM supports the design control requirement of ISO 9000.

The next section covers 5S-CANDO, a set of activities for cleaning, organizing, and arranging the workplace.

5S-CANDO

Rudyard Kipling's poem *The 'Eathen* covers the basic idea of 5S-CANDO quite well.

> The 'eathen in 'is blindness bows down to wood an' stone;
> 'E don't obey no orders unless they is 'is own;
> 'E keeps 'is side-arms awful: 'e leaves 'em all about,
> An' then comes up the regiment an' pokes the 'eathen out.
>
> All along o' dirtiness, all along o' mess,
> All along o' doin' things rather-more-or-less,
> All along of abby-nay, kul, an' hazar-ho,
> Mind you keep your rifle an' yourself jus' so!
>
> abby-nay = "Not now." kul = "Tomorrow." hazar-ho = "Wait a bit."

5S-CANDO is a systematic technique for cleaning, organizing, and arranging a workplace. The acronyms are, respectively:

5S	CANDO
Seiri = Clearing up	Clearing up
Seiton = Organizing	Arranging
Seiso = Cleaning	Neatness
Shitsuke = Discipline	Discipline
Seiketsu = Standardization	Ongoing improvement

5S reduces the friction or muda of searching for tools because every-thing has a place. American manufacturers introduced this practice no later than 1911, and the principles in the following excerpt are as applicable today as they were then:

> In the average shop, these bolts [for clamping work to machines] lie around on the floor: rarely is there a full assortment accessible. Need-ing four-inch [102-mm] bolts, say, the mechanic looks around for them, fails to find a full set, and concludes to use six-inch [152-mm] bolts. Blocking up is necessary and he probably has to screw the nut down an extra inch. Because of the rough care the bolts get, the thread may be damaged, and he has trouble in getting the nut down. In many cases, as motion studies and observations have shown, he consumes from ten to twenty times as many minutes as the clamping ought to take.
>
> Now, each instructions card specifies, in hundredths of an hour, the time allowed for setting the work in the machine. [Notice that scientific management quantified setup time.] Such specification would be useless, of course, unless the proper blocks and bolts were provided for the workman's use. So the planning department sees that a full supply of blocks and bolts of varying lengths are kept in the tool racks. With each job, the mechanic receives the particular size of bolt best suited to the task, just as though these were standard machine tools instead of accessories usually neglected. Further-more, before they are restored to the rack after use, every thread and nut is inspected to make sure they are still in perfect condition. Try any bolt in the tool room and the nut turns easily under your fingers.
>
> Except by comparison of the time consumed in certain oper-ations before and after the reorganization, no conception can be gained of the unbelievable wastes attending some of the less com-mon processes. . . . Now every part has its symbol and its place in the stores room; every operation in assembling has been standard-ized (The System Company 1911, 69–70).

This excerpt shows that:

1. The need to hunt for tools, fixtures, and parts is extremely wasteful.

2. Rough handling and neglect make tools and fixtures harder to use. Preventive maintenance assures that they are easy to use, and this plays a major role in reducing setup time. This in turn supports SMED.

3. American manufacturers introduced arranging (seiton or organizing) long before anyone ever heard of "5S."

Another basic principle of 5S-CANDO is that problems cannot hide in a clean factory. A dirty workplace can conceal defective equipment while, for example, a leak will be very obvious on a clean floor. Ford used light-colored paint to expose dirt instead of hiding it. White or light-gray walls and floors themselves became the inspector's proverbial white glove. Schonberger (1986, 69) shows how cleanliness helps keeps equipment running at Detroit Diesel Allison and Harley-Davidson. Workers realize that they should not only lubricate the machines but also clean them several times daily. Leaks then become obvious instead of hiding the liquid in a pool on the machine or on the floor. The worker can then fix the problem before it results in unplanned downtime.

A clean workplace also helps people work more efficiently. 5S-CANDO even applies to construction work:

> It is a recognized fact that a cluttered-up floor under a workman's feet will tire him quite as much as the productive work that he is doing. A smooth-planked floor will enable a bricklayer to lay many more brick[s] than will earth that has been leveled off (Gilbreth 1911).

5S Elements

1. *Clearing up: "When in doubt, throw it out."*

   - Your wastebasket is your best friend. (A corollary of Murphy's Law says that, for every piece of paper you throw away, two take its place.)

   - E-bay and similar Internet auction sites also are your friends. Serviceable but unwanted tools and equipment—including forklifts, power supplies, grinders, lathes, office equipment including copiers and fax machines, laboratory instruments, bulldozers, and CNC (computer numerical control) machine tools can be sold (or bought) online.

   - Remove nonessential items from the workplace. Put red tags on apparently nonessential items; if no one claims them, store them offline or discard them.

   - Organize tools and equipment by frequency of use. Anything that people use very frequently should remain by the workstation. Intermediate usage items should be near the workstation. Items that see only occasional use should be kept elsewhere. In all cases, see the next step: Arranging.

2. *Arranging: "A place for everything, and everything in its place."* The opposite of, " 'E keeps 'is side-arms awful: 'e leaves 'em all about . . ." Tools, parts, and equipment should all be easy to find. Every tool should have a specific place for storage when it's not in use (Figure 4.4). Many toolboxes, for example socket wrench sets, are organized this way.

3. *Neatness.* Keep everything clean. This eliminates dirt and debris that can interfere with equipment or damage the product. Also, as discussed by Ford and Schonberger, neatness gives leaks and similar malfunctions nowhere to hide.

4. *Discipline: "Mind you keep your rifle an' yourself jus' so!"*

   • Checking and cleaning equipment becomes routine. Standardization can include incorporating cleaning and preventive maintenance schedules into work instructions. Maintenance logs show when these activities are completed.

   • Scheduled preventive maintenance is an aspect of Taylor's scientific management. A "tickler system" should have a portfolio for each day of the year, in which notices could be placed. These notices would ". . . come out at proper intervals throughout the year for inspection of each element of the system and the inspection and overhauling of all standards as well as the examination and repairs at stated intervals of parts of machines, boilers, engines, belts, etc., likely to wear out or give trouble, thus preventing breakdowns and delays" (Taylor 1911a, 117). Today, of course, a computerized planner can handle this task. Scheduled preventive maintenance supports ISO 9000's process control requirement.

5. *Ongoing improvement.* Look for even better ways to organize and clean the workplace. Root out friction and muda, the insidious enemies of productivity that can easily work their way into any job.

More on Preventive Maintenance

The Ford Motor Company used constant preventive maintenance to assure that its machinery kept running smoothly. Taylor described the benefits of preventive maintenance explicitly:

> The machines of the country are still driven by belting. The motor drive, while it is coming, is still in the future. There is not one establishment in one hundred that does not leave the care and tightening of the belts to the judgment of the individual who runs

"Standardized sample tools and methods of storage at the Link Belt factories"

This photo shows machine tools on a rack that has a specific position for each tool: "At the right, notice that the mnemonic symbol for each tool is on a little card above it."

"This corner of a tool room illustrates the axiom of order—a place for everything, everything in its place."

Figure 4.4 Arranging (seiton, or organizing) in the United States, 1911 or earlier.

Source: The System Company, *How Scientific Management Is Applied* (1911).

the machine, although it is well known to all who have given any study to the subject that the most skilled machinist cannot properly tighten a belt without the use of belt clamps fitted with spring balances to properly register the tension. And the writer showed . . . that belts properly cared for according to a standard method by a trained laborer would average twice the pulling power and only a fraction of the interruptions to manufacture [downtime] of those tightened according to the usual methods. *The loss now going on throughout the country from failure to adopt and maintain standards for all small details is simply enormous* (Taylor 1911a, 125–26, emphasis is ours).

While machine shops no longer use belts, of course, the underlying principle still applies. Standardized preventive maintenance improves tool availability (by preventing downtime), efficiency, and/or work quality. Note also the role of calibrated gages and instruments in preventive maintenance. Kipling provides an appropriate conclusion to a discussion of 5S-CANDO:

> Keep away from dirtiness—keep away from mess.
> Don't get into doin' things rather-more-or-less!
> Let's ha' done with abby-nay, kul, an' hazar-ho;
> Mind you keep your rifle an' yourself jus' so!

There are also a number of visual controls that support 5S-CANDO (in terms of workplace arrangement), poka-yoke or error-proofing, JIT manufacturing (through, for example, kanban squares), and jidoka or autonomation (through andon lights).

VISUAL CONTROLS

Caravaggio (in Levinson 1998, 134–35) summarizes visual control systems as follows:

> Visual controls identify waste, abnormalities, or departures from standards. They are easy to use even by people who don't know much about the production area. . . . A visual control system has five aspects:
>
> 1. Communication: Written communications are easily accessible
>
> 2. Visibility: Communication with pictures and signs
>
> 3. Consistency: Every activity uses the same conventions

4. Detection: There are alarms and warnings when abnormalities occur

5. Fail-safing: These activities prevent abnormalities and mistakes

Suzaki (1987, 107–12) compares visual controls to a nervous system, a form of just-in-time information transfer. Wayne Smith (1998, 159) says, "Basically, the intent is to make the status of the operation clearly visible to anyone observing that operation." The recurring theme of synergy cannot be overemphasized. Lean manufacturing and quality improvement techniques are mutually supporting and reinforcing. Visual controls support other activities as shown in Table 4.3.

Gilbreth (1911) provides some excellent examples from the bricklaying trade. These principles apply very broadly.

The stimulating effect of color upon workers is a subject to be investigated by psychologists. The results of their study should be of great benefit, especially to indoor workers. Motions could

Table 4.3   Visual controls and other productivity and quality improvement activities.

Activity	Visual Controls
5S-CANDO (arranging)	Marked positions for tools and materials.
Autonomation (jidoka)	Andon lights, buzzers.
Error-proofing (poka-yoke)	Color-coding or other markings that help assure proper assembly or operation.
JIT manufacturing	Kanban squares, cards, containers (empty container as a signal to make more product), lines on the floor to mark reorder points.
Safety	Colored labels for materials: red for flammable, blue for health hazard, yellow for oxidizer, white for corrosive.
Statistical process control	Control charts must be easily visible to anyone who is associated with the operation. (Charts that can only be viewed on a computer in an engineer's office don't qualify.)
Continuous improvement	A visible production management system (discussed in the next section) should indicate problems that interfere with production goals. Problems require not only containment (immediate actions that restore production) but permanent corrective actions that keep them from returning. The Ford Motor Company's Team Oriented Problem Solving, 8 Disciplines (TOPS-8D) is a systematic approach for doing this.

undoubtedly be made simpler by the proper selection of the color of painting and lighting in the workroom.

In our work we have to deal chiefly with color as a saver of motions. Color can be seen quicker than shape. Therefore, distinguishing things by their color is quicker than distinguishing them by the printing on them. Examples:

- The various pipes in a pipe gallery can best be recognized by painting them different colors. [This reinforces and augments labels that identify pipe contents. Such labeling is a modern workplace safety requirement.]

- The right-hand end of the [brick] packet is painted black, in order that when carried in the right hand of the laborer it can be placed so that the bricklayer can pick up each brick without spinning or flopping the brick in his hand.

- Painting tools different colors, and also the place where they are to be placed in the drawer or the chest the same color, saves motions and time of motions when putting them away and finding them next time. [This supports arrangement, the second element of 5S.]

- When low-priced men bring packages of any kind to higher-priced men to use or handle, the packages should always be painted, stenciled, or labeled with a distinguishing color on one end and on top. This will enable the low-priced workman to place the package in the manner called for on the instruction card with the least thought, delay, and motions. It will also enable the high-priced man to handle the package with no such lost motions as turning the package around or over. [The statement about low- and high-priced workers refers to the fact that unskilled workers transported materials to skilled workers and trade workers such as masons.]

- Oftentimes the workmen who are best fitted physically for their work cannot read, or at least cannot read English. [This is less likely to be a consideration today.] Even if they could, it would take some time to read the stenciled directions on the non-stooping scaffold [a scaffold that was designed specifically to make bricklaying easier] to the effect that "this side goes against the brick wall." It will greatly reduce the number of motions to paint the side that goes next to the wall a different color from the side that goes away from the wall.

Schonberger (1982, 145) describes how Japanese firms color-code part containers to reduce the time that workers need to find the right parts. Caveat: a small percentage of the population is color-blind. It may be worth coding objects with patterns as well as shapes, a well-known practice for traffic controls. The red stop light is always on top of or to the left of the other lights. Stop signs are always octagonal and no other traffic sign has this shape. Yield signs are similarly triangular.

Visible Management and Production Control

Wayne Smith (1998, 159) lists some qualifications for a visible management system. Each work area should have a display board. It can be manual or electronic but its information *must* remain current. It should show:

1. *What the operation is trying to make.* The measurement is the takt rate, or desired production per unit time.

2. *What the operation is achieving.*

3. *What problems, if any, are hindering achievement of the production goal?* Who is responsible for fixing the problem? When will it be fixed?

The following excerpt from Frederick G. Coburn's "Laying Out Work for Each Man" recognizes these considerations explicitly and provides techniques for addressing them:

> For each constituent operation of an order an instruction card corresponding to the standard order is written at the time the work is planned, and, when issued to the workman, *it is hung in plain sight in a tin rack at the workman's bench.* To insure definitely the complete occupation of the employee's time three jobs are assigned him; he is working on one, the second is ready— all materials and appliances at hand, and the third is either ready or the stock is in the material or the milling department. When a job is completed, the mechanic hangs his card on a hook on the lower right hand corner, moves up the other two, and goes on with his work. . . . *The rack always shows the foreman what the man is doing, and calls attention to the jobs ahead, so that it is of the very greatest value in coordinating the work of the various departments* (The System Company 1911, 83, emphasis is ours).

The system that Coburn describes incorporated another advantage of modern just-in-time systems. Neither defects nor upstream stoppages can hide in piles of inventory; the next workstation discovers them very quickly:

> . . . it is also necessary that each man get through with his job in time for the next man to take it up, for that next man isn't going to be caught loafing if he values his job. [This phrasing reflects management attitudes of the early 20th century but it also shows the element of *pull*. Stoppage of a downstream operation due to lack of work leads to rapid inquiries to the upstream operation.] And the work must be done right, or the next man will kick [complain], lest the boss find him with a piece of imperfect work. *The whole thing becomes an interlocking and smoothly working mechanism if correctly planned and supervised: and the most trouble occurs under the conditions producing apparently the smoothest running under the old system* (The System Company 1911, 84, emphasis is ours).

The last part of this excerpt suggests that scientific management practitioners recognized that problems can hide in non-lean, inventory-fat factories. The system appears to run smoothly because inventory decouples operations from one another. If a worker gets a defective piece, he or she can simply discard it (or throw it back on the pile) and work on another. He might not bother to notify the operation that produced the defect. This is not an option in an inventory-lean factory.

Schonberger (1982, 2, 25) raises the same point. If a workstation makes a defective part in a lean factory, the downstream internal customer will complain immediately because there is no inventory buffer to conceal the defect. Schonberger also adds (32, 78, 91, 114–115) that the Japanese reduce buffer stocks and take workers out of the line to *expose problems*. Removal of the inventory reveals quality problems and unsteady production rates. The Japanese also reduce work-in-process (WIP) by removing kanban, without which upstream stations cannot generate WIP. Removal of workers reveals inefficient methods and equipment that can hide when extra labor is available.

Visible demand, a related concept, ties in with production control. The idea is that each workstation should be able to determine its downstream customer's needs (W. Smith 1998, 204). Kanban systems can use various signals as shown previously.

Visual Controls and Quality

Schonberger (1982, 57) says that Japanese inspectors want visual and obvious quality indicators at every process. These must be easy for anyone in the shop to understand, and they should not require interpretation by an engineer or technician. Ford (1926, 77) describes an electrical camshaft timing tester that embodies this principle. Flashing lights indicated a defect, and the operator would then examine an index on a handwheel to identify the faulty cam.

Visual controls make it easy for everyone in the factory to see production status and quality problems. Color-coding of labels supports workplace safety. Color- and shape-coding parts supports error-proofing (poka-yoke), which this chapter discusses in more detail on pages 60–63.

The next section covers cycle time reduction, a prerequisite for making to order and not to unreliable market forecasts. Short cycle times equip a company to become a just-in-time supplier.

CYCLE TIME REDUCTION

Lean manufacturing offers another key benefit in cycle time reduction. Cycle time is the time between a job's start and its completion. Smith (1998) defines cycle time as:

$$\frac{\text{Inventory (units)}}{\text{Demand}\left(\dfrac{\text{units}}{\text{time}}\right)}$$

This does not necessarily mean that simply removing inventory will reduce cycle time. Per Standard and Davis (1999), inventory is often the flower, not the root, of all evil. It is a symptom of something else: batch-and-queue operations that introduce waiting time, overproduction, variation in processing times, or other problems. Removal of the inventory without correction of the underlying problems may simply reduce demand (or *through-put*, a key performance measurement under the theory of constraints). The factory will make fewer units but it will take as long to get them through.

To illustrate this concept, restate the equation for cycle time as: Inventory = Demand (or throughput rate) × Cycle time. That is, inventory is a function of cycle time (the effect), not the other way around (the cause). Smith (1998, 40) again reinforces Standard and Davis' point that inventory is the flower, not the root, of all evil:

1. Inventory is present because of an underlying problem that has not been corrected

2. Directives or other attempts to remove the inventory will not work unless the underlying problem is corrected

3. Removal of the underlying problem, however, causes the inventory to vanish by itself

Lead time is the time between an order's placement and its delivery. If the factory builds parts to order, as it should, lead time should be cycle time plus order processing time. The latter should be negligible, especially in an electronic commerce environment, but bureaucracy can make it significant.

Harley-Davidson, which now has product lines other than motorcycles, manages to deliver a certain product so quickly that it receives payment for it before it has to pay for the raw materials. It is almost like finishing before one starts. This is the basic idea of lead time reduction.

Shorter lead times also improve responsiveness to customer needs and simplify production planning. Instead of manufacturing to fulfill unreliable market forecasts, the factory can manufacture to order. Adams Citrus Nursery in Haines City, Florida, built a continuous flow greenhouse. Seedlings traveled in Citripots that moved slowly on a rail system through the greenhouse. Mature trees emerged several months later. The lead time to grow a tree fell from three years to nine months, for a 75 percent reduction. Adams therefore did not have to guess what kind of trees customers would want two or three years in advance. It could grow them to meet the current year's market requirements. Adams became the world's largest citrus tree supplier and even gained a market in Saudi Arabia (Standard and Davis 1999, 2–3).

Details of Adams' process are not known but the following educated guess was provided:[5] the company may well use sunlamps or ultraviolet lamps to accelerate production. This sounds expensive until one considers the cost of producing trees that cannot be sold because the market forecast was wrong, and the inventory carrying costs for a 3-year growth time. Furthermore, a faster growth rate increases the nursery's capacity. *The idea that cheap is often dear is a recurring theme in lean manufacturing.* Beware of false economies.

Imai (1997, 22–23) raises a key idea that will be worth repeating later. "There is far too much muda [waste, often wasted time] between the value-adding moments. We should seek to realize a series of processes in which we can concentrate on each value-adding process—Bang! Bang! Bang!— and eliminate intervening downtime." Standard and Davis (1999, 63) show the concept graphically, as in Figure 4.5.

Smith (1998, 10) defines manufacturing cycle efficiency (MCE) as follows:

$$MCE = \frac{Value\text{-}adding\ time}{Total\ cycle\ time}$$

He adds that *MCE can often be less than 1 percent.*

The idea is that the work really takes a small fraction of the time the job spends in the factory. This is a vital concept behind single-minute exchange of die (SMED), which this chapter discusses in more detail on

The Process: Four Operations

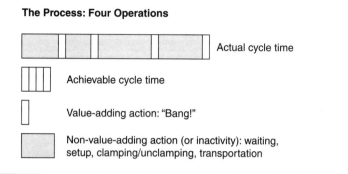

Actual cycle time

Achievable cycle time

Value-adding action: "Bang!"

Non-value-adding action (or inactivity): waiting, setup, clamping/unclamping, transportation

Figure 4.5 Enormous opportunities for lead time reduction.

pages 72–77. Standard and Davis (1999, 61) use golf, a sport to which the Japanese can certainly relate, as an example. The golf club head actually contacts the ball for about 0.02 seconds; this is Imai's "Bang!" or value-adding moment. One can easily relate this to, for example, a metal-forming die or punch striking the workpiece. Even the golf club's swing is not part of the value-adding instant, and a machine tool adds value only when it *contacts* the workpiece.

Suppose that a game takes four hours and 90 strokes. Only 1.8 seconds, or 0.0125 percent of the time, is actually spent on the value-adding "Bang!" Pedestrianism is a healthy aspect of a golf game but Henry Ford defined it (in the context of walking to get tools and parts) as a poorly-paid line of work, as did his contemporaries:

> In another factory, because the tools were inconveniently arranged, the employer was paying sixty per cent of his total pay-roll for time spent in moving about to get things. He thought he was hiring men for work; he was in fact paying for pedestrian endurance. And everybody lost on the deal (Basset 1919, 71).

The same applies to materials and work in process, and the golf analogy shows how to think about cycle time. How can we reduce the time between the value-adding moments?

Keep the Work Moving

Figure 4.6 shows why work should be kept in continuous motion. The idea is not to keep people and equipment busy (a dysfunctional measurement that often results in the production of unusable inventory) but to keep work moving. There are two operations, the first of which requires 0.5 hour. The second requires 1 hour.

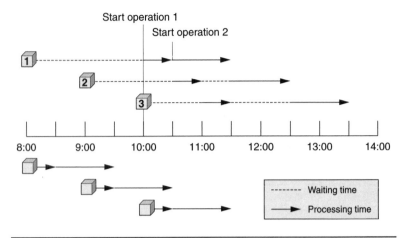

Figure 4.6   Keep it moving.

The top of the figure shows a situation in which work is allowed to accumulate at the first operation, which does not begin processing until 10:00. The sins, or rather the waiting times, of the first operation are visited on the second. A queue of three parts is allowed to accumulate at the first operation before anything happens. Then the second and third jobs have to wait (0.5 and 1 hour respectively) because the workstation is busy. They arrive at the second workstation at 0.5 hour intervals but, since that station requires an hour, the second and third jobs end up waiting there as well. The cycle time for each job is 3.5 hours: 1.5 for processing (the achievable minimum), 1 as obvious waste in waiting for the first operation, and another hour waiting for busy equipment.

The bottom shows the case in which work is processed as soon as it is available, and there is no waiting time at all.

Cycle Time and "Factory Physics"

Standard and Davis (1999) use the term *factory physics* to quantify cycle time as a function of (1) variation in processing times and (2) equipment utilization. We will see later that Henry Ford and other scientific management practitioners often tried to subdivide factory operations into elemental tasks. One worker, for example, would place a screw while another would tighten it. The obvious effect is to eliminate the non-value-adding action of picking up and putting down a screwdriver each time. Another effect is to *suppress variation in the task's time*. This may be why the Ford production control system worked as well as it did even though it was not a true pull system like kanban or drum-buffer-rope (DBR).

If σ is the standard deviation of the time for a given task whose average time is t, the coefficient of variation is

$$cv = \frac{\sigma}{t}$$

Suppose that a complete job at an autonomous workstation has an average time of eight minutes with a coefficient of variation of 0.125. The job then has a standard deviation of one minute. Now assume that this job can be broken into four two-minute operations that have the same coefficient of variation.

- The standard deviation for each operation is therefore 0.25 minutes

- The standard deviation of a sum is

$$\sigma_{\sum X_i} = \sqrt{\sum \sigma_i^2}$$

where σ_i is the standard deviation of the ith X value

- In this case the work still takes 4×2 minutes = 8 minutes but the standard deviation is

$$\sqrt{4 \times \left(\frac{1}{4}\right)^2} = \frac{1}{2} \text{ minute}$$

This reduces the variation in process time and thus suppresses the problem that Goldratt and Cox describe in *The Goal:* the tendency of supposedly balanced production lines to accumulate inventory. Standard and Davis (1999, 107) go on to state that *the average inventory in a factory is proportional to the standard deviation of the cycle time.* It does not depend on the mean cycle time, only on the variation. Acceleration of processes (and reduction of waiting) can therefore reduce cycle time but not inventory.

Delays in Systems with Spare Capacity: "Hurry Up and Wait."

Standard and Davis (1999, 234–235) provide a compelling argument against seeking 100 percent equipment utilization—a common performance measurement in far too many factories. They also show why the balanced production line simulation in Goldratt and Cox (1992) leads to steadily-increasing inventory levels. That simulation consisted of a series of workstations that could process one to six (as determined by a die roll) units per turn. Each station had an average capacity of 3.5 units per turn so the system was perfectly balanced. When it was tested, however, it accumulated inventory.

Standard and Davis present the following equation for cycle time in queue for any given workstation:

$$CT_q = \left(\frac{c_a^2 + c_e^2}{2} \right) \left(\frac{u}{1-u} \right) t_e \qquad (4.1)$$

CT_q = cycle time in queue (waiting for the workstation)

$$c_a = \frac{\sigma_a}{t_a} = \text{coeffecient of variation for arrivals at the station}$$

$$c_e = \frac{\sigma_e}{t_e} = \text{coeffecient of variation for effective processing time}$$

$$u = \frac{t_a}{t_e} = \text{utilization}$$

t_e = effective processing time

This suggests, of course, that *cycle time in queue becomes infinite at 100 percent equipment utilization.* Wayne Smith (1998, 189–90) also shows that problems occur as utilization approaches 100 percent. Furthermore, variation in effective processing time t_e and arrival time t_a—a function of processing time and arrival time variations from previous workstations—increases cycle time.

Standard and Davis add that equation 4.1 is exact when the arrival times follow the exponential distribution (that is, arrivals per unit time follow the Poisson distribution). Equation 4.1 seems to predict far too high waiting times when there is no variation in the arrival times and the effective processing time follows a normal distribution.[6] This discussion's primary purpose is not, however, to identify a particular equation for average waiting times but rather to illustrate the effect of both variation and utilization. The effect is described best as "hurry up and wait."

Consider, for example, a workstation with a processing time of, on average, 19 seconds. Production control releases a new job every 20 seconds. One would expect no buildup of inventory at the workstation because it has excess capacity. Here is what actually happens, though, if there is variation in the processing time: The tool can finish the job early or late. Suppose it finishes in 17 seconds instead of 19. It has supposedly gained two seconds but no work is available; the tool simply gains two seconds of idle time. This is the "hurry up" part.

Suppose it finishes the second job in 21 seconds. The third job has to wait one second; it accumulates an extra second of cycle time. This is the "wait" part. The average of 17 and 21 seconds is, incidentally, 19 seconds. The problem is that the third job feels only the "wait" effect of the variation in processing time. *This is why parts can accumulate cycle time in queue even in a system that has excess capacity.* Despite the $u/(1 - u)$ factor in equation 4.1, variation in processing and arrival times due to the "hurry up and wait" effect is probably a far greater problem than high utilization.

Ford said he wanted to design jobs to give each worker all the time he needed but not a second more, a statement that sounds terrifying in the context of equation 4.1. It's easy to visualize rows of warehouses and piles of inventory between workstations. Charlie Chaplin's performance of assembly line work in *Modern Times* seems like a realistic portrayal of how this system should have worked, or rather not worked. The Ford system had at least two features, however, that mitigated against the effects of equation 4.1.

1. Ford said specifically, "Use work slides or some other form of carrier so that when a workman completes his operation, he drops the part always in the same place—which place must always be the most convenient place to his hand—and if possible have gravity carry the part to the next workman for his operation" (Ford 1922, 80). The upstream station delivered its work as soon as it was completed, thus removing any variation in delivery time from σ_a in equation 4.1. There was still, of course, variation in the upstream station's own processing time, which translates into variation in arrival time for the next station. However:

2. Ford wanted to make the work as automatic as possible. All the worker should have to do is (possibly) load the piece, pull a lever or push a button, and remove the completed unit. Completely automatic processes have very little variation in effective processing time (σ_e) and this may be another reason why the Ford production system worked as well as it did. Many semiconductor manufacturing operations also have very little variation in the time that silicon wafers actually spend in a workstation.

In summary, the Ford system *achieved very high equipment utilization in a balanced production line* by *suppressing variation in arrival and processing times.* This ties in closely with the concept of takt times, and modern factories should pay close attention to these concepts.

The next section discusses single-minute exchange of die, a method for reducing non-value-adding setup time. It applies to almost anything, not only machine tools that actually use die.

SINGLE-MINUTE EXCHANGE OF DIE

Single-minute exchange of die (SMED) is, like most other well-known productivity and quality improvement techniques, American in origin.

> *In a certain shop with which we are familiar a piece had to have several holes of different sizes drilled in it, a jig being provided to locate the holes. The drills and the sockets for them were given to the workman in a tote box. The time study of this job revealed several interesting facts. First, after the piece was drilled the machine was stopped, and time was lost while the workman removed the piece from the jig and substituted a new one. This was remedied by providing a second jig in which the piece was placed while another piece was being drilled in the first jig, the finished one being removed after the second jig had been placed in the machine and drilling started.*
>
> —Robert Thurston Kent, in Gilbreth (1911)

This statement leaves absolutely no doubt that American scientific management practitioners identified the issue of internal setup (setup that requires stoppage of the tool) versus external setup (setup that a worker can perform while the tool is running).

Figure 4.7 distinguishes non-value-adding setup work from value-adding work very explicitly. It is an outstanding illustration of the thought process behind SMED.

The identification and reduction of non-value-adding setup operations is the foundation of what we now know as single-minute exchange of die, of which Japan's Shigeo Shingo is the best-known exponent.

Single-minute exchange of die allows conversion of batch-and-queue operations to single-piece or small-lot operations and can reduce lead times. The key to understanding SMED is, *know when the operation adds value to the product or service.* Recall Masaaki Imai's observation:

> There is far too much muda [waste, often wasted time] between the value-adding moments. We should seek to realize a series of processes in which we can concentrate on each value-adding process—Bang! Bang! Bang!—and eliminate intervening downtime (Imai 1997, 22–23).

SHOP MANAGEMENT   171

THE MIDVALE STEEL CO.

Form D—124.   Machine Shop.......................18.........

ESTIMATES FOR WORK ON LATHES

OPERATIONS CONNECTED WITH PREPARING TO MACHINE WORK ON LATHES AND WITH REMOVING WORK TO FLOOR AFTER IT HAS BEEN MACHINED		NAME ....
OPERATIONS	TIME IN MINUTES	Sketch .... Number........... Order............... Weight........ Metal................. Heat No. .... Tensile Strength.... Chem. Comp...... Per cent. of Stretch .... HARDNESS, Class...............

| Putting chain on, Work on Floor |
| Putting chain on, Work on Centers |
| Taking off chain, Work on Floor |
| Taking off chain, Work on Centers |
| Putting on Carrier |
| Taking off " |
| Lifting Work to Shears |
| Getting Work on Centers |
| Lifting Work from Centers to Floor |
| Turning Work, end for end |
| Adjusting Soda Water |
| Stamping |
| Center-punching |
| Trying Trueness with Chalk |
| "   with Calipers |
| "   with Gauge |
| Putting in Mandrel |
| Taking out " |
| Putting in Plug Centers |
| Taking out "   " |
| Putting in False Centers |
| Taking out "   " |
| Putting on Spiders |
| Taking off " |
| Putting on Follow Rest |
| Taking off " |
| Putting on Face Plate |
| Taking off "   " |
| Putting on Chuck |
| Taking off " |
| Laying out |
| Changing Tools |
| Putting in Packing |
| Cut to Cut |
| Learning what is to be done |
| Considering how to Clamp |
| Oiling up |
| Cleaning Machine |
| Changing Time Notes |
| Changing Tools at Tool Room |
| Shifting Work |
| Putting on Former |
| Taking off " |
| Adjusting Feed |
| "   Speed |
| "   Poppet Head |
| "   Screw Cutting Gear |

SIGNED   TOTAL

OPERATIONS CONNECTED WITH MACHINING WORK ON LATHES

OPERATIONS	Speed	Feed	Cut	Tool	Inches	Minutes
Turning Feed In						
" " "						
" Hand Feed						
" " "						
Boring Feed In						
" " "						
" Hand Feed						
" " "						
Starting Cut						
" "						
Finishing Cut						
" "						
Fillet						
"						
Collar						
"						
Facing						
"						
Slicing						
"						
Nicking						
"						
Centering						
"						
Filling						
"						
Using Emery Cloth						
" " "						
TOTAL						

Machining — Two Heads Used
"   — One Head Used
Hand Work
Additional Allowance

TOTAL TIME
HIGH RATE
LOW RATE

Remarks

Time actually taken

FIGURE 6. — INSTRUCTION CARD FOR LATHE WORK

The left-hand column identifies non-value-adding setup activities explicitly and the right-hand column identifies value-adding machining actions. The thought process that goes with this form is as valid today as it was more than a century ago. The activities in the left-hand column do not transform the product so they add no value to it. SMED reduces the proportion of these non-value-adding activities.

Figure 4.7   Non-value-adding setup work versus value-adding machining work.
Source: Frederick W. Taylor, *Shop Management* (1911): 171.

As with many other techniques and methods, military necessity drove the development of a version of SMED. The value-adding process in question was a literal "bang!"—a volley from musket-armed infantry. The wooden and then the paper-wrapped musket cartridge evolved to externalize the setup operation of measuring out a charge from a powder horn. The breechloading firearm got another non-value-adding setup operation, use of the ramrod, out of the process. The modern practitioner should think the same way and recognize that the actual value-adding moment may indeed be very brief. In a machine shop, it takes place only while the tool is in contact with the workpiece. A drill takes tangible time to go through a part but a punch's value-adding moment is a literal "bang," a fraction of a second.

Taiichi Ohno (1988, 39) shows how the need for SMED arose at Toyota when the company moved from producing large runs of single products to small runs of diverse products. This required frequent die changes, so setups had to be rapid. Ohno adds that in the 1940s die changes at Toyota often required two to three hours. By the late 1960s they were down to a few minutes.

External and Internal Setup

"Exchange of die" is deceptive because it suggests that the technique applies only to machine tools that use die. *The idea is to perform as much setup work as possible while the machine is working on another job.* This is *external setup*, or setup that is done outside the machine. *Internal setup* requires the machine to shut down.

Exchanging die on a machine tool is a non-value-adding setup activity, and so is exchanging lithographic masks in a semiconductor exposure tool. The latter is like a giant camera whose sophisticated optics often bring its cost into the millions of dollars. A silicon wafer with a photosensitive coating plays the role of the film. A mask with a microscopic wiring pattern is the photographic subject. The idea is to reproduce the wiring pattern on the wafer.

Other semiconductor processing equipment uses gas bottles or metal sputtering targets. Replacement of materials is a setup operation. Loading wafers, especially batches of them, into a process tool is setup, as is removing them. In all cases, external setup loses no processing time while internal setup does.

A plastic vacuum molding operation does not involve machine tool dies, but SMED still applies to it (Robinson 1990, 322–23). The steps are as follows:

1. Join a moveable mold to a fixed mold

2. Pump the air out of the mold to create a vacuum

3. Inject resin

4. Open the mold and take out the finished part

The process adds value when it injects the resin to form the part. Evacuating the mold, while necessary, is a non-value-adding internal setup step. It's hard to see how to make it external; the mold must be closed before the air can be removed. Shigeo Shingo recommended the purchase of a tank whose volume is 1000 times that of the mold. The vacuum pump could decompress this tank while the mold was making a part. In step 2, the tank would be connected to the closed mold; opening a valve would remove 99.9 percent of the air instantly. The pump would still have to remove the remaining air if the process required a high vacuum. The change, however, converts much of the internal setup to external setup: something that can be done while the tool is working.

Robinson (1990, 334–37) shows how it is often possible to secure or loosen a bolt or screw with a single turn. Consider a bolt that secures a die to a press. If the nut has fifteen threads, the worker must turn it fifteen times. Only one turn, the one that tightens the bolt, adds value. The other fourteen turns are waste. A change that makes the setup require only one turn converts the bolt into a *functional clamp*.

Pear-shaped holes in flanges allow the flange to be placed over the bolts that hold it down. The bolt heads go through the large openings of the pear-shaped holes. A slight turn puts the narrower parts of the pear-shaped openings under the bolt heads, and single turns clamp the flange into place (Figure 4.8). It's equally simple to remove the flange, as a single turn on each bolt is enough to loosen it.

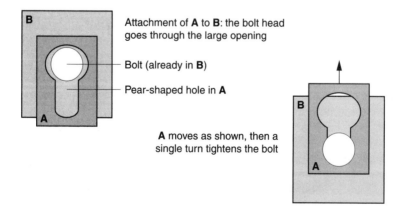

Figure 4.8   Pear-shaped holes allow rapid clamping and unclamping.

This shows that it's possible to tighten a bolt with a single turn. It's even possible to tighten one with a one-sixth (60-degree) turn in a way that seats several threads for extra strength. The concept behind the split-thread bolt (Robinson 1990, 337) may well have come from a 19th century invention, the interrupted screw artillery breech. We have already seen that military necessity spurs the development of methods (for example, error-proofing of communications) and technology (making musket setup external via the cartridge with its premeasured powder charge) that only later find their way into civilian applications. The ability to shoot more quickly at a target that shot back was doubtlessly an incentive that does not exist in manufacturing enterprises. Nineteenth century artillery developers knew what Shigeo Shingo said perhaps a century later: one can't afford to waste time in setup, and turning a screw several times to seal an artillery breech took too much time.

The answer was the interrupted screw: a breechblock with threads on only half its diameter. The gun bore had a corresponding set of threads that were offset so the gunner could simply insert the block into place. A quarter (90-degree) turn then sealed the breech. The closure is so strong that it can withstand the pressure in a sixteen-inch (40.6-cm) gun. The split-thread bolt (Figure 4.9) works the same way and a one-sixth (60-degree) turn will tighten it.

SMED sounds impressive because it can improve the capacity of any workstation that involves setup. The theory of constraints shows, however, that *SMED is futile for increasing overall factory capacity except when it is*

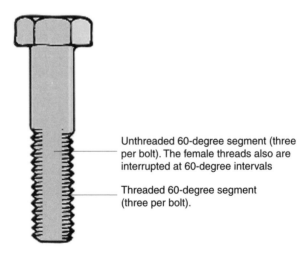

Unthreaded 60-degree segment (three per bolt). The female threads also are interrupted at 60-degree intervals

Threaded 60-degree segment (three per bolt).

Figure 4.9   Split-thread bolt.

applied at the constraint! SMED can, however, reduce or eliminate the need for batching at non-constraint operations. Although this does not increase capacity it reduces cycle times and promotes single-piece or small-lot flows. While defects and errors are never desirable anywhere, the theory of constraints shows where to focus poka-yoke and in-line inspection[7] activities.

This section has illustrated SMED, the purpose of which is to reduce non-value-adding setup time. Imai's description is very useful: the value-adding work often takes a fraction of a second. In some cases, it is a literal "bang!" as a punch or hammer strikes the workpiece. SMED externalizes the work that surrounds this value-adding moment.

Poka-yoke is another technique that is commonly associated with Dr. Shigeo Shingo. Like SMED, however, error-proofing (poka-yoke) originated in the United States.

ERROR-PROOFING (POKA-YOKE)

Error-proofing is another lean manufacturing technique that also supports ISO 9000. Its basic premise is that anything that requires human intervention and judgment to prevent mistakes is a mistake waiting to happen. Dr. Shigeo Shingo introduced this technique to Japan as *baka-yoke* (fool-proofing). He changed it to poka-yoke because workers inferred from baka-yoke that management perceived them as stupid (Shingo 1986, 45).

Here is an example of poka-yoke. "While the welding operation is in progress, fan-shaped plates, operated by cams, cover in turn all operating buttons except the one needed for the next move. It is impossible for the operator to go wrong" (Ford 1930, 198). Gilbreth's (1911) advice to color-code objects to facilitate proper orientation also is a form of error-proofing. Color-coded wires and matching connection points are an example of this, but keep in mind that some people are color-blind.

Another error-proofing technique is designing keys into parts to prevent improper assembly. The large and small prongs on a polarized electrical plug, with matching openings in the electrical outlet, are an example. It is impossible to insert the plug backward.

Gages and automatic sorters that prevent the use of substandard parts also are a form of error-proofing. Nonconforming raw materials or components that enter the constraint can cause downtime, scrap, or rework.[8] It is therefore very worthwhile to keep such items out of the constraint. Downstream incorporation of bad parts into good units from the constraint also can be devastating, since post-constraint scrap is irreplaceable.

Here is an example of a device that culls nonconforming bushings from a manufacturing process before they can get into the final product. ". . . the undersized [bushings] go into one chute and the good ones into another. Those which remain on top are dropped from the end of the rollers into the over-sized chute" (Ford 1926, 76). That is, the conforming and oversized bushings will not fit into the first opening, through which the undersized units drop. The oversized ones do not fit through the second, through which the conforming ones drop. This sorter was adjustable in 0.1 mil (2.54 micron) increments. Robinson (1990, 249) cites a stem tightener sorter that works the same way. These are examples of what Shigeo Shingo calls *self-check systems.*

Robinson (1990, 284) describes how a process for cutting grooves under bolt heads would sometimes produce uncut bolts because of chucking problems on the cutting tool. It cost only $75 to build a device (Figure 4.10) that stops bolt heads with uncut grooves from progressing. An alarm sounds if the device does catch a defective bolt, thus allowing the factory personnel to fix the chucking problem.

The concept of error-proofing also applies to workplace safety. The basic principle at Ford's River Rouge plant was, "can't is better than don't." That is, set up the equipment and the job so workers *can't* injure themselves instead of telling them, for example, *"Don't* monkey with the buzz saw." The System Company (1911a, 114) cites the latter instruction as "one of New England's colloquial proverbs, to which too many four-fingered men call attention." Even Henry Ford's production chief, Charles Sorensen, lost two fingertips when he made wooden patterns for iron casting molds. Interlocks, guards over moving parts, and lockout-tagout are examples of accident-proofing.

This section has covered error-proofing (poka-yoke) and related concepts like self-check systems. Error-proofing supports ISO 9000 and QS-9000 requirements. Our next subject, TOPS-8D, is a systematic and extremely versatile problem-solving technique. While not specifically a lean manufacturing method, it is an important supporting technique, and also has excellent synergy with ISO 9000's corrective and preventive action requirement.

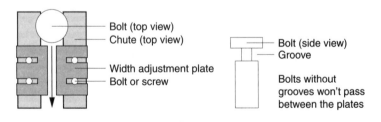

Figure 4.10   Error-proofing a chute for grooved bolts.

TEAM ORIENTED PROBLEM SOLVING, 8 DISCIPLINES

Ford Motor Company's Team Oriented Problem Solving, 8 Disciplines (TOPS-8D) technique is not specific to lean manufacturing but it is so useful that it deserves mention. It strongly supports the corrective action requirement of ISO 9000. Its original form is reactive; a company performs an 8D to fix a quality problem. Most proactive productivity and quality improvement activities, however, especially those that involve self-directed work teams, can be organized and documented as 8Ds.

8D is more versatile than Six Sigma's RDMAICSI (Recognize, Define, Measure, Analyze, Improve, Control, Standardize, and Integrate) because it applies, with little or no modification, to a wider range of improvement activities. Six Sigma has a quantitative focus but 8D requires only a clear quantitative or qualitative definition of the problem. This ties in with the simple definition of friction for frontline workers: *If it's frustrating, a chronic annoyance, or a chronic inefficiency, it's friction.* Workers should be able to define friction the way someone once defined obscenity: "I know it when I see it." It is always nice to quantify the effects of improvements but don't let this consideration block constructive action. Recall Ross Perot's advice: the first person who sees a snake kills it, and the same applies to friction.

8D's foundation, like that of most other improvement activities, is the Plan–Do–Check–Act (PDCA) improvement cycle (Juran and Gryna 1988, 10.25–10.26):

1. *Plan:* Identify improvement opportunities as gaps between existing and desired performance. Plan an experiment.

2. *Do:* Make the change and collect data.

3. *Check:* Assess the results. Observe the change's effects.

4. *Act:* Go to the "plan" step of a process control cycle to hold the gains.

   - Define internal customer requirements and output requirements and measurements. Set up feedback process control (for example, statistical process control, automatic process control) where appropriate. Define supplier requirements. Plan the work; process sheets are an example.

   - We can add to the above, "Standardize the improvement by incorporating it into work instructions. Deploy it as a best practice by looking for other activities to which it might apply."

Repeat the cycle to get further improvement. In the following 8D approach, disciplines 1, 2, 4, and 5 fall roughly under "plan." The sixth includes "do" and "check," and the seventh is "act." The eight disciplines are:

1. Use a team approach. The team will ideally include frontline workers who are intimately familiar with the operation in question. This is an aspect of "plan" in PDCA.

    • The team should include a *project champion:* a person with the resources and authority to implement the change or solution. Kaizen Blitz, a proactive technique for improving productivity, also calls for a project champion. This reinforces the argument that 8D can serve as a framework for seemingly different improvement programs.

2. Develop a working definition for the problem. This also is the "plan" in PDCA.

    • Juran and Gryna (1988, 22.35) emphasize that the problem definition must be clear and unambiguous. Beware of words with multiple meanings.

    • "Problem" means that something is not performing according to existing standards. A proactive 8D will, if successful, elevate the existing standard: replace "problem" with "improvement opportunity."

3. Contain the problem. This means segregating nonconforming products so they don't reach customers, shutting down malfunctioning manufacturing equipment, and so on.

    • Containment is primarily reactive and it does not play a major role in improvement activities.

    • Containment does not really fall into PDCA; it is something we must do to protect the customer.

    • Recall General LeMay's admonition, "Stop swatting flies and go after the manure pile." Containment is swatting flies. We may have to do it to protect our customer but we will never be rid of the flies unless we finish the 8D process.

4. Identify the problem's root cause. Certain aspects of root cause analysis constitute planning because they examine the existing process. Design of experiments can, however, be a miniature PDCA cycle of its own.

- Tools include traditional quality improvement tools like check sheets, Pareto charts, and cause-and-effect (fishbone) diagrams

- Design of experiments is useful for testing proposed improvements, for example, experimental versus control or plan-of-record

5. Select a permanent correction for the root cause. Selection of the change also is planning under PDCA.

6. Carry out the permanent correction ("do") and verify its effectiveness ("check").

7. Prevent the problem from returning by making the change a permanent part of the activity.

   - Revise work instructions or operating instructions and other standards

   - In 8D's proactive mode, standardize the improvement to make it permanent and hold the gains ("act")

   - Addition: see if the change can be deployed elsewhere as a best practice

8. Recognize the team's accomplishments.

   - Public recognition shows the entire organization that the improvement technique works, and this encourages widespread use. It is an important part of change management.

Figure 4.11 is an 8D flowchart from Fairchild Semiconductor's built-in quality (BIQ) initiative.

The "Question to the Void" or "Five Whys"

TOPS-8D includes a technique called the "question to the void." It means asking *why?* until the problem's root cause is exposed. Taiichi Ohno (1988, 17) gives an example of using the "Five Whys" process to discover the reason for a machine stoppage.

1. The machine stops because of overloads, which blow the fuse. (A purely reactive response, one that, to use General LeMay's metaphor, 'swats flies instead of going after the manure pile,' would be to replace the fuse.)

2. Overloads occur because of inadequate bearing lubrication.

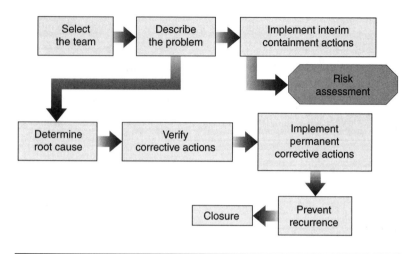

Figure 4.11 Flowchart.

3. Inadequate lubrication is caused by insufficient flow from the lubrication pump.

4. The cause of the insufficient flow is a worn-out lubrication pump shaft.

5. The shaft wears out because absence of a strainer for solids allows entry of metal scrap.

Assuring that the strainer is always attached would, again paraphrasing General LeMay, get rid of the manure pile—the underlying root cause of the trouble. Other measures like routine cleaning of the strainer to prevent blockage also come to mind.

Then there is the series of whys that leads back to the beginning. Readers of Goldratt's and Cox's *The Goal* (1992) will appreciate this one.

"Why can't we get orders through our production line quickly?
"Our inventory often contains obsolete or defective parts."
"Why are the parts obsolete or defective?"
"They become obsolete when we make them on a just-in-case basis, no one uses them, and an engineering change makes them obsolete. If an operation makes defective parts and they go to inventory instead of the next operation, no one notices the problem until the parts come out of inventory."

"Why do we make parts on a just-in-case basis and put them into inventory?"
"We want a ready supply of anything we might need to get an order through our production line quickly."

A series of whys that takes us back to the starting point shows that we obviously have a problem. Inventory is to this factory what the whiskey bottle is to the heavy drinker: not the solution but the problem.[9]

Quality Improvement Process

Fairchild Semiconductor's quality improvement process (QIP) is an eight-step system for proactive improvement (Figure 4.12). It roughly follows the 8D structure; there is, of course, no need for a containment step because the process is not doing anything wrong. This underscores the difference between *fixing* something and *improving* it. A problem, like the production of defective pieces, requires correction (containment plus removal of the underlying root cause). A process that is making no defective parts can still be improved, and this is the thrust of lean manufacturing.

This section has provided an overview of the eight-discipline (8D) problem-solving process, which is adaptable to a wide variety of problems

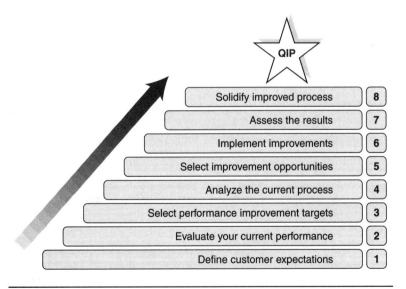

Figure 4.12 Quality improvement process.

and even to proactive improvements. Fairchild Semiconductor's Mountain-top plant uses 8D to fulfill much of the ISO 9000 preventive and corrective action requirement. Fairchild's quality improvement process is a related procedure for proactive improvement.

SIX SIGMA

Six Sigma refers not only to process capability,[10] but to the quality and productivity improvement program as described by Harry and Schroeder (2000). Harry and Schroeder add quality improvement and problem-solving tools to the Six Sigma umbrella. These activities rely on the standard PDCA improvement cycle, or its rough equivalent, Define, Measure, Analyze, Improve, Control (DMAIC).

Harry and Schroeder extend DMAIC to an eight-step process: recognize, define, measure, analyze, improve, control, standardize, and integrate. This is roughly parallel to Ford Motor Company's TOPS-8D approach. TOPS-8D is actually more versatile because of Six Sigma's focus on data and process variables. TOPS-8D can address these issues and many others as well. TOPS-8D uses cross-functional teams (an aspect of the kaizen blitz), looks for root causes, and implements and tests permanent corrections or improvements.

Six Sigma roles include:

- *Executive leadership*, which is the prerequisite for any quality or productivity improvement initiative.

- The *champion,* whose role is wider than that of an 8D champion. Like the latter, Six Sigma champions control resources that might be necessary for an improvement's implementation. They also identify projects, perform benchmarking and gap analysis, create operational visions, and provide managerial and technical leadership.

  - *Deployment champions* are high-level personnel whose roles are similar to those of the CEO, president, or executive vice president.

  - *Project champions* work at the business unit level. They might include site management team leaders.

- *Black Belts,* who apply Six Sigma techniques (mostly traditional problem-solving and improvement techniques, including statistical methods such as design of experiments) to individual projects.

They also provide mentoring, teaching, and coaching, and they look for improvement opportunities inside and outside the organization. Master Black Belts train Black Belts, design cross-functional experiments, facilitate meetings, and collect and organize information. Both are organizational change agents.

- *Green Belts* require less training than Black Belts. They learn basic statistical and other problem-solving tools that they can apply to their jobs, and they also participate in Six Sigma projects.

Six Sigma also calls for best practice deployment. "Six Sigma is about more than successfully completing individual projects. . . . the organization develops a set of best practice procedures and standardizes them" (Harry and Schroeder 2000, 114). This agrees with Taylor's (1911) guidance on this subject. Six Sigma is therefore not a particularly new or innovative program but it encompasses many well-tested and generally accepted practices.

Six Sigma Process Capability

The words "six sigma" refer to six standard deviations between the process mean and the specification limits (Figure 4.13).

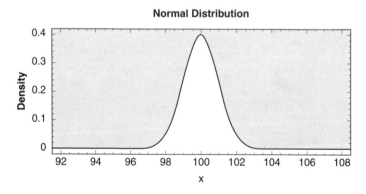

The tail area above 106 (the upper specification limit)
contains 1 billionth of the distribution's density,
as does the region below 94 (the lower specification limit).

Figure 4.13   A Six Sigma process. Specification limits are [94, 106], μ = 100, σ = 1.

If this process remains in control, no more than one piece per billion will exceed either specification. Under a worst-case assumption, if the process drifts 1.5 standard deviations, no more than 3.4 pieces per million will be defective (Figure 4.14). Six Sigma terminology is "3.4 defects per million opportunities (DPMO)."

Six Sigma ties in with process capability. Process capability indices measure the process' ability to meet specifications. The capability indices[11] are ratios of the specification width, or the distance between one specification limit and the process mean, to the process variation. Lower variation yields higher capability indices. Those for an in-control Six Sigma process are 2.0, while the generally accepted minimum for industrial processes is 1.33.

This ties in with design for manufacture (DFM) and ISO 9000's design control requirement by requiring designers to account for the capability of manufacturing equipment. Designers must work with the shop floor (cross-functional teams) to know the capabilities of the tools there. As an example (using figures 4.13 and 4.14), the designer should know that the process' standard deviation is one unit. The specification width should therefore be at least twelve units to achieve Six Sigma process capability.

Beware, however, of manufacturing processes that don't follow a normal (bell curve) distribution (Levinson, 2000a). Calculation of the nonconforming fraction as a function of the number of standard deviations (sigmas) between the process mean and the specification relies on the assumption of a normal distribution. Violation of this assumption can lead to orders of magnitude of error, for example 1000 ppm defective where one expects 10 ppm.

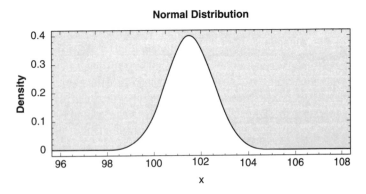

The tail area above 106 (the upper specification limit)
contains 3.4 millionths of the distribution's density.
The region below 94 is negligible.

Figure 4.14   Six Sigma process with 1.5 sigma shift. Specifications: [94, 106].

One can always compute the nonconforming fraction by fitting the *actual* distribution. The lognormal distribution is very straightforward; the logs of the measurement follow the normal distribution. Other common distributions include the gamma and Weibull distributions. A one-sided specification limit (for example, maximum for impurities or particle counts, minimum for strength) is a warning that the distribution *may* be nonnormal. This is especially true when use of the normal distribution suggests a significant chance of getting impossible measurements like negative impurity levels. Always look at the process' data histogram and normal probability plot, never assume that the data are normal. In summary, Six Sigma process capability assures infinitesimal defect rates only when the process is normally-distributed and, of course, in control.

Six Sigma uses tested and well-established productivity and quality improvement tools. It can and has brought about substantial improvements in applications to which it is suited. It does not, however, consist of anything particularly new and its RDMAICSI approach is not as versatile as TOPS-8D.

KAIZEN BLITZ ("LIGHTNING CONTINUOUS IMPROVEMENT")

Cox and Blackstone (1998) define *kaizen blitz* as:

A rapid improvement of a limited process area, for example, a production cell. Part of the improvement team consists of workers in that area. The objectives are to use innovative thinking to eliminate non-value-added work and to immediately implement the changes within a week or less. Ownership of the improvement by the area work team and the development of the team's problem-solving skills are an additional benefit.

Laraia, Moody, and Hall (1999, 172–73) give additional details and the following basic ideas:

Think of kaizen as a means of concentrating resources and driving improvement to unheard-of levels. Kaizen is *an ideal vehicle for capitalizing on the skills that organizations and individuals in the organization already possess*—for bringing these skills to action for *immediate effect*.

The aim of initial projects, after all, is to *encourage the organization to further action* by achieving *quick, highly visible, and sustainable success*. How better to bring this about than by employing a few simple tools that people are familiar and comfortable with (emphasis is ours).

The underlying concept of kaizen blitz's role in change management and transformation of organizational culture is far from new. Taylor identified the role of visible results in transforming culture:

[The workers'] real instruction, however, must come through a series of object lessons. They must be convinced that a great increase in speed is possible by seeing here and there a man among him increase his pace and double or treble his output. . . . It is only with these object lessons in plain sight that the new theories can be made to stick (Taylor 1911a, 132–33).

There is, in fact, evidence that kaizen blitzes or their close equivalents took place under Taylor's original scientific management:

In one plant the men have themselves re-designed nearly every machine in the place, with astounding results in the way of production, quality, and lowering of sales price, with an increase of wages to the men and profits to the company. In another factory, *within six months from the time the workers were given a voice in the management, they devised more improved machinery than had been known in that particular industry within twenty years* (Basset 1919, 133, emphasis is ours).

Basset also cites (105–6) another shop in which the workers posted a sign that proclaimed everyone an "efficiency engineer." The factory doubled its output without adding more workers or equipment.

This is a strong argument in favor of self-directed work teams (SDWTs, see Wentz, "Teaming to Win," and Sands, "Zero Scrap Actions," in Levinson, 1998) like those in place at Fairchild Semiconductor's plant in Mountaintop, Pennsylvania. These teams divide responsibility among members much as professional societies and service organizations do. There are, for example, rotating assignments for safety, quality, production, and leadership (coordination of meetings). Autonomous work groups are similar. Worker empowerment requires training in team dynamics and manufacturing technology, and training must include lean principles.

Kaizen blitz adds an emphasis on worker participation to Taylor's scientific management. When the frontline workers improve productivity themselves, cultural change occurs far more quickly. In both systems, however, visible results promote cultural transformation.

Here are the key principles:

1. *Kaizen blitz is not a new program.* It uses existing productivity and quality improvement techniques such as:

   • Work teams

- 5S-CANDO

- Factory layout changes and work cell redesign

- Reduction of all forms of waste (muda, friction)

- Pull manufacturing systems (kanban, synchronous flow manufacturing)

- Value analysis (value-adding versus non-value-adding work)

- Single-minute exchange of die

- Transition from batch to small-lot or single-piece flow

2. The focus is on achieving rapid (within a week) and identifiable results.

   - Action basis: don't bother with much planning in offices or conference rooms, go out to the factory floor and do it

   - Use marker boards in the work area so everyone can see the tools and proposed rearrangement plans at the same time[12]

   - Instant gratification builds morale and encourages further use of the techniques

   - Workers learn improvement techniques by using them

3. Kaizen blitz is also a change management tool that transforms the work culture.

   - Demonstrable results overcome the "we've always done it that way" paradigm

An ideal first project should qualify as a textbook case study. It should have clearcut objectives and high visibility, and it should fill a business need. It should involve frontline workers and encourage them to apply their knowledge of the job. The roles of improvement techniques like 5S-CANDO, layout rearrangement, and inventory reduction should be obvious.

Hold the gains through standardization by making the new methods permanent. Document new methods in a work instruction or operating instruction. Use best practice deployment to apply the new knowledge in other areas where it is appropriate.

Kaizen blitz requires workers to understand the friction concept, and the frontline worker is indeed often in the best position to recognize friction. *Kaizen blitz is a powerful change management tool because it delivers tangible results very quickly.*

Rearrangement of tools is a common element of a kaizen blitz. The next section will show how the factory layout can affect productivity.

FACTORY LAYOUT STRATEGY

Heizer and Render (1991, 390) cite the following possible factory layouts.

1. *Fixed-position,* in which the product (usually a large engineering project like a ship, highway, or a building) does not move. Workers and equipment must organize themselves around the product. This is not of major interest in lean manufacturing and the chapter will not devote a subsection to it.

   • Assembly of a car in one place is an example. This is how Henry Ford made cars during the very early 1900s.

   • A shipyard is characteristic of a fixed-position layout. Ward (1999, 36–37) describes how destroyer escort construction was accelerated during the Second World War by prefabricating large components in Denver, Colorado and sending them to Mare Island in California for final assembly. Innovative techniques such as this deserve mention because this kind of thinking does promote lean production.

   • It's obvious to everybody (this phrase is a warning sign) that a ship must be constructed right side up. It was actually far easier for the welders to work with the deck floors above them and the ceilings below them. When the work was complete, it took three minutes for a steel cradle to turn the ship over. Remember a concept that was discussed earlier: *don't take any aspect of the job as a given.*

2. *Process-oriented* or *job-shop* layout, in which departments are organized by machine types. It is most suitable for *low-volume* and *high-variety* production.

   • If the factory makes a huge variety of parts (and a goal of lean manufacturing is to reduce the quantity of part numbers) that do not relate well under group technology, this layout may be necessary. Unfortunately, it introduces transportation costs automatically and these are waste by definition.

3. *Cellular manufacturing* layout, which Heizer and Render call a subset of process-oriented layout. Work cells are, however, organized around part families and not around machine types. This is more characteristic of the product-oriented layout, and perhaps cellular manufacturing is best described as a combination of both. The work cell plays a critical role in lean manufacturing environments.

   • Work cells cut transportation distances greatly and improve work flow.

   • The authors cite R. E. Flanders (1925) as the originator of the work cell.[13] The concept, however, appears explicitly in Ford (1922).

4. *Product-oriented* layout, which is most suitable for continuous and repetitive production.

   • The stereotypical assembly line, which includes fixed-position machines or "monuments" with conveyors between them, falls into this category. It often makes only one product, or one product with some customized features.

Sorensen (1956, 164) identified the advantages of single-floor factories and applied this layout to Ford Motor Company's River Rouge foundry. The idea was to reduce the number of conveyors and elevators, and also the waste of labor time in going up and down.

Process-Oriented Layout (or Job-Shop Layout)

Another heading for this section might be, "Don't arrange your factory this way unless you have no other choice." The process-oriented or "farm" layout puts all machines of the same kind—drill presses, grinders, lathes, mills, and so on—into departments.

Semiconductor plants also use this layout. There are separate departments for spin coaters, lithographic exposure tools, wet chemical etchers, ion implanters, and diffusion furnaces. The utility-intensive nature of semiconductor processing equipment probably plays a major role in dictating this layout. Placement of hydrofluoric acid etching stations in different parts of the plant would, for example, require multiple sets of special plumbing to remove this dangerous chemical. The exposure tools require vibration isolation and this may be easier to achieve if they are all in one place. Finally, the capital-intensive nature of the process and the product's light weight makes interdepartmental transportation costs less important.

Figure 4.15 shows that this layout increases transportation distances (waste) and handling (more waste). A *spaghetti diagram* is often useful for mapping work flows in factories (Laraia, Moody, and Hall 1999, 176–78).

Heizer and Render (1991, 394) cite the following advantages and disadvantages of the process-oriented layout. In its favor it is very flexible because jobs can be processed on any machine. A work cell must, in contrast, have at least one of every necessary tool. Downtime on one machine in a job shop will not stop production because another machine of the same type can do the work. Downtime on a machine in a work cell can stop the entire cell. Under the theory of constraints, however, this is not a major problem unless the downtime is at the constraint operation.

The process-oriented layout's disadvantages include generally slower production times, complicated job scheduling, inventory, and of course transportation costs. *It is also most suitable for the low-volume*

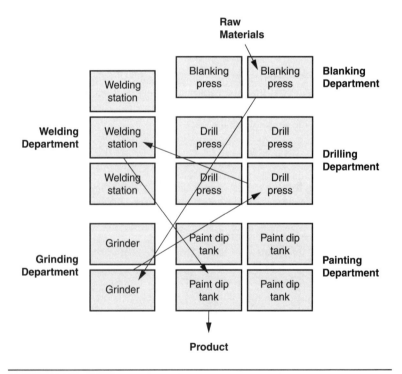

Figure 4.15 Spaghetti diagram of a process-oriented layout.

and high-variety production that is characteristic of job shops. The factory can produce many different products that probably don't fall into convenient group-technology families. Each product or (small) product group requires a different set and/or sequence of operations, which inhibits work cell organization.

The best process-oriented layout minimizes the cost of transporting materials and subassemblies between departments.

$$\text{Cost} = \sum_{i=1}^{n} \sum_{i=1}^{n} X_{ij} C_{ij}$$

where X_{ij} is the number of loads to move between departments i and j and C_{ij} is the cost of the movement. This cost is, of course, one hundred percent waste (muda) under the lean manufacturing definition.

There are scientific techniques for optimizing a process-oriented layout but they do not always yield the best result. CRAFT (computerized relative allocation of facilities technique) seeks to minimize the total material handling costs. There are, however, $n!$ (n factorial) potential layouts for n departments so even computers do not always yield optimum solutions. A change in production requirements and thus interdepartmental transportation (the X_{ij} s) can make a formerly optimum layout obsolete.

Heizer and Render (1991, 400) cite SPACECRAFT, a computer program that optimizes the layout for a multistory factory.

Cellular Manufacturing and Group Technology

Arrangement of the machines in order of production operations reduces non-value-adding transportation time and often inventory. The process-oriented job shop uses hand trucks, forklifts, and carts to move work between tools. This encourages movement in batches or lots instead of single units, an issue that recurs on a larger scale as "less than load" (LTL) in freight management. When the workstations are in the same order as the operations, workers can transfer single pieces by hand, conveyor, or work slide. American manufacturers recognized this principle prior to 1911:

> Only by relating each machine with the others in such a way that production will follow in straight lines without confusion, can the highest economy of operation be attained.
>
> . . . Figure I: In this machine shop of the Mueller Machine Tool Company, floorspace is well utilized by arranging the machines logically with respect to production (The System Company 1911a, 124–25).

Henry Ford described cellular manufacturing in 1922:

We started assembling a motor car in a single factory. Then as we began to make parts, we began to departmentalize so that each department would do only one thing. As the factory is now organized each department makes only a single part or assembles a part. *The part comes into it as raw material or as a casting, goes through the sequence of machines and heat treatments, or whatever may be required, and leaves that department finished* (Ford 1922, 83–84, emphasis is ours).

Notice that the department is organized around the *part*, not the machine type. This is the basis of cellular manufacturing.

Many manufacturing processes must make different products. The flexibility to switch between products quickly facilitates small-lot or single unit production and it provides a competitive advantage. *Cellular manufacturing* and *group technology* are important and mutually supporting techniques.

Cubberly and Bakerjian (1989, 5–14) define group technology as a design for manufacture (DFM) technique. "Group technology (GT) is an approach to design and manufacturing that seeks to reduce manufacturing system information content by identifying and exploiting the sameness or similarity of parts based on their geometrical shape and/or similarities in their production process." Group technology not only facilitates manufacturing, it makes the designers' jobs easier.

Furthermore, the editors state,

The grouping of related parts into part families is the key to group technology implementation. The family of parts concept not only provides the information necessary to design individual parts in an incremental or modular manner, but also provides information for rationalizing process planning and forming the machine groups or cells that process the designated part family.

Group technology means assigning similar part numbers (the fewer the better) to *production families*. All members of a production family go through the same manufacturing operations, for example blanking, grinding, drilling, welding, and painting (Schonberger 1986, 10). A manufacturing cell can be laid out to accommodate this work flow and it can make any member of the product family (Figure 4.16).

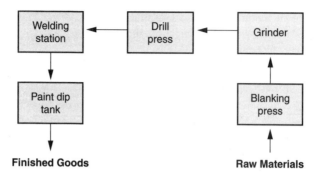

Figure 4.16   Manufacturing cell.

Some key considerations:

- Counterclockwise, right-to-left work flow often results in better ergonomics.

- Distances between work stations are as small as possible so people don't have to transport parts. No inventory is kept between workstations (except before a constraint if there is one).

- The cell can make any part number in the part family. It is acceptable to include a part number that skips one or more operations in the cell.

Unitary Machines

The unitary machine takes the work cell concept to the next step: a single machine performs all the operations necessary for making a part. It reduces handling (non-value-added activity with opportunities for handling damage), job lead time, and often cost. Per Ford (1930, 131–32), operations could be combined by using the principle of the turret lathe. Drilling, boring, reaming, turning, and facing operations that would have been performed on eight or nine different machines could be done on a single machine. There was only one loading and one unloading activity instead of seven or eight. This reduced the opportunities for handling damage. "Thus there is no lost time or motion; the parts are handled but twice—once in loading as a rough forging and again in removing as a finished part." Ford adds that this approach could reduce cycle times from hours or days to minutes, while doubling accuracy and reducing costs by fifty percent.

Notice the unitary machine's potential for enormous reduction of cycle time. Mark Leeser, general manager of Hitachi Seiki USA in Congers, New York, echoes this idea (Olexa 2001):

> The trend in manufacturing today is to get the part done with one machine tool, which decreases the required floor space. It also reduces the amount of labor needed to produce the part because one person can tend several machines. By keeping the workpiece in one machine, you get a more accurate part, because it's being machined in one setup, which also eliminates the labor to move the part. By eliminating refixturing or regripping the part, additional fixtures for different machines don't need to be purchased, and you don't have the lead time and expense involved in building and deploying them.

Numerically-controlled machines are modern examples of this technology. Schonberger (1986, 115) says that unitary organization works best when:

1. The machine is close to preceding or subsequent operations

2. There is more than one unitary machine, which provides backup and flexibility

3. Setups and changeovers are rapid and simple (SMED applies here)

4. The unitary machine should be portable

5. The unitary machine must be reliable

Group Technology and Fixed-Position Equipment

Process industries often use equipment that often cannot be easily moved (W. Smith 1998, 161). Examples include distillation towers and absorbers, which are effectively fixed structures. It may also be difficult to rearrange smaller semiconductor manufacturing equipment because of all the utility connections. Smith points out that group technology still allows the factory to minimize the number of *routes* that a particular product or product family can take. Each route becomes the equivalent of a work cell that processes a group of related products.

Another problem with the process-oriented layout is that supposedly identical equipment really isn't. Smith adds that, if there are 24 possible routes for a product, the factory is effectively making 24 different products. Consider a semiconductor product that can go through any of three metallization tools, four furnaces, and two ion implanters. There are effectively

24 different ways to make this product. This complicates statistical process control,[14] and it's why there is often substantial between-lot variation in semiconductor products.

Smith, who is writing from a chemical process industry (DuPont) perspective, recommends the dedication of specific routes to create "superhighways for critical products." The equipment on the dedicated routes doesn't have any starts, stops, or changeovers. This also has the potential to reduce process variation and improve process capability.

This approach is subject to practical capacity considerations. The factory cannot dedicate a constraint operation to a specific product without losing overall capacity. Nonetheless, the concept allows factories with process-oriented layouts to gain some benefits of group technology.

Autonomous Assembly

Autonomous assembly is a similar concept for humans, in that one worker or a team of workers performs a complete and identifiable job. Instead of assembling two engine components, for example, a team builds an entire engine.

Intrinsic motivation makes the task its own reward and promotes sustained intensity of effort. Lack of intrinsic motivation was identified as a problem as early as 1919:

> . . . in the factory job he probably does not know what he is doing; he is simply going through a monotonous routine and doing certain work because he is told to do it, and without an idea of just what part he plays in the final product, what his relative importance is, or what is the value of that with which he is working. He has no measure of personal responsibility in the factory as high as that of the machine he operates. There is nothing to draw out of him the natural and fundamental instinct of creation (Basset 1919, 80).

Benefits of autonomous assembly include the following elements of intrinsic motivation, the absence of which is noticeable in the quoted excerpt.

1. *Task identity:* workers can see the finished piece, not a "widget" the purpose of which is not apparent from its appearance

2. Task significance: workers understand why the product is important to the organization

3. Autonomy

4. Skill variety: workers exercise different skills

Autonomous assembly has drawbacks, however, when workers have to make non-value-adding motions. Exchanging one tool for another, or putting down a tool and picking up a part, are non-value-adding *setup* operations for humans. Henry Ford used autonomous assembly during his company's early days but he and his associates discovered quickly that task subdivision made jobs far more efficient. The person who tightened a nut was not the same person who placed it on the bolt—a practice that Charlie Chaplin parodied in *Modern Times*. To tighten the bolt, the person who had placed it would have had to pick up a wrench, and "pick up" is a non-value-adding activity. Then he would have had to put it down to place the next nut. Breaking a large job, like assembly of an engine, into very simple activities improved its efficiency immensely. Also, picking up and putting down the same tool dozens or hundreds of times a day is no more intrinsically motivating than performing a single activity, it may be far more tiring, *and it does not pay very well*. The company, at any rate, found ways to make machines do routine tasks like placing and tightening nuts.

Schonberger (1986, 116) cites similar experiences by other companies: autonomous assembly results in long cycle times. Nowadays machines can do most of the dull and repetitive work and employees can use their brains to run the machines. Modern planners should look at both efficiency and intrinsic motivation.

Product-Oriented Layout

Heizer and Render (1991) state, "Product-oriented layouts are organized around a product or a family of similar high-volume, low-variety products." This is the opposite of the situation for a process-oriented layout, which is best for low-volume and high-variety products. Similarity implies the application of group technology.

The authors also discuss *fabrication lines*, that build components, and *assembly lines*, that assemble the components. It is easy to envision work cells that make parts that go to an assembly line. The underlying idea is parallel instead of sequential assembly.

The product-oriented layout works best when:

1. The product is high-volume.

2. Demand for the product is sufficiently stable to justify investment in specialized equipment (and/or immovable "monuments").

3. Standardization of the product justifies the investments described above. (Recall that group technology is a technique for standardizing parts so one series of operations can make an entire part family.)

4. Raw materials and subcomponents are sufficiently uniform to work reliably in the standardized equipment. Henry Ford used automatic sorting devices to exclude anything that didn't meet this requirement (an example of what Shigeo Shingo would later call a self-check system).

This section has covered different factory layouts and their advantages and disadvantages. Remember that transportation is a non-value-adding activity that creates opportunities for handling damage. Transportation invites batching, and this increases cycle times.

THE WORK ENVIRONMENT

We have already seen that a job's tools, methods, and materials should not be taken as givens; neither should the environment. The work environment itself can promote or hinder productivity. Frank B. Gilbreth had several observations on this subject. Heating, ventilation, and air conditioning (HVAC); lighting; and even music and entertainment all can influence productivity. Remember that ISO 9000 asks whether the work environment is suitable for the job. Temperature and humidity also affect measurement capability and gage calibration, and these are also ISO 9000 considerations.

Heating, Ventilation, and Air Conditioning

The needs of the product or process may dictate temperature and humidity (T&H) requirements. Even if they do not, a comfortable workplace can improve productivity:

> Maintaining desired temperature in summer as well as winter by forcing into workrooms air that has been passed over heating or refrigerating coils has a great effect on the workman. Many factories, such as chocolate factories, have found that cooling the air for better results to the manufacturing process also enables the workers to produce more output—an output quite out of proportion to the cost of providing the air (Gilbreth 1911).

The System Company (1911a, 33) gives an example in which a company's directors objected to the cost of installing fans in a sewing factory. The factory's manager proved not only that the fans were not expensive, they were cheap. Production increased sixteen percent after their installation. Furthermore:

> Good lighting, heat and ventilation pay. A factory manager by painting the walls cream-white toned up his whole working force. Hard to measure, he could "feel" the improvement brought about by this simple means (The System Company 1911a, 77).

Lighting

Lighting can influence both quality and productivity. ISO 9000 even asks about the suitability of lighting. This is yet another example of synergy between lean manufacturing and ISO 9000. Although the famous Hawthorne study showed that experiments with lighting improved productivity simply because the workers received attention from the experimenters, Gilbreth cited some real considerations that are equally applicable today:

> The subject of lighting has, indirectly as well as directly, a great influence upon output and motions, as upon the comfort of the eye depends, to a large extent, the comfort of the whole body. The arrangement of lighting in the average office, factory, or house is generally determined by putting in the least light necessary in order that the one who determined the location of the light may be able to see perfectly. This is wrong. The best light is the cheapest. By that is not meant that which gives the brightest light. In fact, the light itself is but a small part of the question. Go into any factory and examine every light, and you will notice that as a rule they are obviously wrong. A light to be right must pass five tests:
>
> a. It must furnish the user sufficient light so that he can see
>
> b. It must be so placed that it does not cause the user's eyes to change the size of the diaphragm when ordinarily using the light
>
> c. It must be steady
>
> d. There shall not be any polished surfaces in its vicinity that will reflect an unnecessary bright spot anywhere that can be seen by the eyes of the worker
>
> e. It must be protected so that it does not shine in the eyes of some other worker
>
> The use of polished brass and nickel should be abandoned wherever it will shine in the worker's eye.
>
> For work done on a flat surface, like the work of a bookkeeper or a reader, the light should be placed where the glare will reflect least in the worker's eyes; where the work is like the examining of single threads, the relative color and figured pattern of the background, as well as good light, is important.
>
> This is obvious. So is nearly everything else in good management. Go into the buildings among the workers, the students, and the scientists and see how rarely it is considered. All of this is not

a question of getting the most out of the light. Light in a factory is the cheapest thing there is. It is wholly a question of fatigue of the worker. The best lighting conditions will reduce the percentage of time required for rest for overcoming fatigue. *The difference between the cost of the best lighting and the poorest is nothing compared with the saving in money due to decreased time for rest period due to less fatigued eyes* (Gilbreth 1911, emphasis is ours).

Manufacturers of that era agreed:

Good light to work by is an investment too infrequently made in the factory. In comparison with the cost of labor, the cost of artificial light is trifling, but there are thousands of skilled mechanics who lose efficiency because of insufficient light (The System Company 1911a, 106).

The writer adds that dirty lamps gave as little as 20 percent of the light that clean ones provided.

Fluorescent lighting may often be a false economy despite its inexpensive operation. Fluorescent lights do *not* emit white light. This is why photographs of factories and offices often have a greenish tint unless the photographer uses the correct filter (FLD) for daylight film. One can look at a fluorescent light through a prism and see several bands whose positions depend on the gas inside the tube. Genuine white light, the kind with which the human eye is designed to work, has a continuous rainbow spectrum. Textile mills recognized this issue more than 90 years ago:

In textile mills this [electric arc] method of lighting is almost essential since color values are distinguishable most exactly by the arc light. Its spectrum is nearest that of the sun (The System Company 1911a, 105).

Our conclusion is that provision of comfortable working conditions and appropriate lighting is cheap—in comparison to the consequences of not providing them. Proper lighting is often a requirement for inspections and quality control. Temperature and humidity control are furthermore not only prerequisites for worker comfort, they are often mandatory for the proper operation of gages. These are ISO 9000 considerations.

INDUSTRY-SPECIFIC TECHNIQUES

This section will illustrate specific methods for making jobs more efficient. These methods often improve quality as well by eliminating built-in variation (common or random cause variation) and defects. They can often save money and reduce lead times by actually removing operations from the route sheet or routing.[15] *The purpose of our examples is to reinforce the universal thought process behind them, which can then be extended to other industries.*

A key point is that *these examples cover what value analysis would define as value-adding activities.* Cost of quality (COQ) analysis would define them as required, as opposed to prevention, appraisal, or failure (rework or scrap). Remember that Ford and Shingo said that most waste is well-disguised. After all, if the waste was obvious someone would have done something about it. Words like "move," "store," "handle," and "sort" imply waste. Phrases like "straighten the part" and "tap the brick into the mortar" imply value-adding or required actions but plenty of waste can hide in them. We will begin with heat treatment, a subject that comes up very frequently and usually in unpleasant contexts—for example, "Herbie [constraint] Number Two" in Goldratt and Cox's (1992) *The Goal.*[16]

Heat Treatment

Heat treatment was why some of the first interchangeable parts weren't. Samuel Colt, in conjunction with Francis Pratt and Amos Whitney, introduced the American System in which interchangeable parts would supposedly eliminate the need for hand-fitting. (In the European System, parts were handcrafted individually and then fitted.) Mid-19th century machine tools could work parts only when they were soft. The parts required heat treatment to harden them after machining and this distorted them unpredictably. Hand-fitting was then required and parts from the same lot could not be interchanged after fitting (Womack and Jones 1996, 153 and note 4).

The Ford Motor Company discovered the problems of heat treatment in the early 20th century. Car axles did not cool uniformly after heat treatment so they had to be straightened afterward. It is easy to imagine the straightening operation becoming an accepted part of the job. Under the COQ model it's not prevention or appraisal (inspection) so a first impression is that it's required. Another perspective, however, is that it's 100 percent rework (failure). Remember that a characteristic of friction is that people can work around it, and it often becomes an accepted part of the job.

Ford's workers recognized that the straightening operation was indeed a form of rework that added no value to the product. The company developed a centrifugal hardening machine immersed the axle shafts in a caustic solution and spun them. This cooled them evenly and almost instantly. Ford eliminated the straightening operation, thus saving $36 million (in 1920s dollars) over four years.

Chapter 6 (single-unit flow) mentions the experience of Omark, a Canadian chainsaw manufacturer. Its heat-treatment operation did not cause problems like the ones Colt and Ford encountered. The problem was a large oven that required large batches of parts. The company considered smaller ovens that could handle small batches, and lasers that could treat single pieces. It eventually found a supplier who could provide a steel alloy that did not require heat treatment. This was apparently even better than the other options it considered because it eliminated the entire heat-treatment operation.

Kalpakjian (1984, 231) suggests another possibility (besides lasers and smaller ovens) for avoiding batch-and-queue heat treatment ovens. Induction hardening can process individual metal parts in seconds. The technique uses an electric field to heat the entire workpiece or sections of it:

> Heat depth can be limited to just the surfaces or can include the entire cross-section. . . . Important advantages of induction heating include increased production, reduced costs, and improved products. Cycle times are a matter of seconds and machines can be completely automated. Precise controls reduce or eliminate scrap, distortion is minimized, and the need for more costly alloy steels is sometimes eliminated (Cubberly and Bakerjian 1989, 41-18–41-19).

The induction heaters also require less floor space than ovens and they can be placed almost anywhere in the factory. This makes them easy to integrate into work cells.

Cubberly and Bakerjian (1989, 41-18–41-19) also cite high-frequency resistance hardening, which can harden specific areas on the surfaces of workpieces. If this is suitable for the product, it reduces energy requirements, avoids workpiece distortion, and shortens cycle times. There is little need to let the part cool afterward because the intense heat is localized. Hardening results from self-quenching of the heated stripe by the surrounding cold metal.

Flame hardening is another surface-hardening technique that can handle small lots. It is suitable for pieces that are too large for induction hardening. The authors also discuss laser hardening, which Omark considered, and electron-beam hardening.

Bricklaying

The book has already covered Frank Gilbreth's observation that placement of the bricks at waist level helps the mason work more quickly without expending more energy. Another apparently required or value-adding step in this trade was to tap each brick with the trowel's handle to get the right thickness for the joint. However:

> Mr. Gilbreth found that by tempering the mortar just right, the bricks could be readily bedded to the proper depth by a downward pressure of the hand with which they are laid. He insisted that his mortar mixers should give special attention to tempering the mortar, and so save the time consumed in tapping the brick (Taylor 1911, 39).

Recall that Omark eliminated heat treatment completely by buying an alloy that did not require it. The bricklaying example reinforces the principle that selection of the right material, even if it is more expensive than the existing one, can save money and improve efficiencies by eliminating operations (Table 4.4).

A Brick Is a Brick?

Gilbreth provides another example of material selection, or rather material specification:

> The most advantageous size of unit to use is a difficult problem to solve, and is often controlled by some outside factor. For example, the most economical size of brick has been determined by the cost and other conditions relating to the making and baking, and not by the conditions of handling and laying. When the conditions of laying are studied scientifically, as they are to-day, one is forced to the conclusion that, for the greatest economy, the size of common brick should be changed materially from that of the present practice in America (Gilbreth 1911).

This statement shows how easy it is to overlook waste while looking straight at it. The brick is taken as a given and the job is designed around

Table 4.4 Elimination of operations.

Example	Action	Operation Eliminated
Ford	Installation of a centrifugal hardening machine	Axle shaft straightening
Omark	Selection of an alloy that does not require heat treatment	Heat treatment
Gilbreth	Specification of properly-tempered mortar	Tapping bricks to achieve the proper joint thickness

the brick. Consider two extremes. At one end are bricks that, although each fills considerable surface area in the wall, are so heavy that the mason tires quickly from working with them. At the other end are bricks that are so small that it takes very little effort to lift one, but each fills so little surface area that the wall grows very slowly. There is, somewhere in between, a brick that fills as much area as possible without being so heavy that it fatigues the bricklayer.

A key goal of the early scientific management practitioners was, in fact, to identify the ideal brick—or, more specifically, the ideal load.

A Shovel Is a Shovel?

It depends on what one is shoveling. A shovelful of iron ore weighs a lot more than a shovelful of ash. At Bethlehem Steel, however, workers often brought their own shovels and used them for different kinds of work. Taylor (1911, 32) observed that people would go from shoveling iron ore at 30 pounds (13.6 kg) per load to slippery rice coal at 4 pounds (1.8 kg) per load. The former was excessively tiring, the latter was very inefficient. Taylor found that the ideal weight for a shovel load was 21 pounds (9.5 kg). This required the availability of 8 to 10 different kinds of shovels. Again, *always* ask whether the job's tools and materials make the work easy and efficient. *Never* take anything, whether a tool, method, or material, as a given.

ENDNOTES

1. The following discussion of the Kyoto Protocol does not necessarily represent Fairchild Semiconductor's position, if any, on this subject. The global warming controversy provides an opportunity to compare lean thinking—the idea of cutting waste and getting the most out of every resource—with approaches that do little more than throw money at problems.
2. The Carnot cycle—which cannot, incidentally, be achieved by any real-world engine—defines the maximum efficiency η of a process that transfers heat from a hot (T_{hot}) reservoir to a cold (T_{cold}) one.

$$\eta = \frac{T_{hot} - T_{cold}}{T_{hot}}$$

   T is absolute temperature (Rankine or Kelvin). Real-world boiler-turbine-condenser-pump power cycles (for example, the Rankine cycle) are less favorable than the Carnot cycle. This also shows that electric heat is extremely inefficient because it returns only a fraction of the original fuel's heat.
3. http://www.ananova.com/news/story/sm_299143.html as of 8/6/01. The tax proposal was seen elsewhere as well.

4. See NFPA 481, Standard for the Production, Processing, Handling, and Storage of Titanium. See also http://www.timet.com/fab-p21.htm (Titanium Metals Corporation, as of 5/16/01) and the International Titanium Organization at www.titanium.org

5. Frank Wheatley, Fellow of IEEE, inventor of the insulated gate bipolar transistor (IGBT), is currently a consultant for Fairchild Semiconductor's plant at Mountaintop, Pennsylvania.

6. The normal distribution is probably not the best model for processing times but it should be adequate. The exponential distribution is definitely out because it suggests the possibility of a zero *processing* time.

7. Shigeo Shingo uses the term "inspection," which is slightly misleading. Quality practitioners know that inspection is among the least efficient quality assurance methods. Shingo's term encompasses automatic sorters and other automatic equipment that does not rely on human judgment.

8. The discussion of the theory of constraints will expand on this considerably. Nonconforming units that waste the constraint operation's capacity cause irrecoverable opportunity costs that the cost accounting system often does not recognize.

9. Antoine de Saint-Exupery's *The Little Prince* has a scene in which a man drinks to forget that he is ashamed because he drinks.

10. Process capability indices measure a manufacturing operation's ability to meet specifications consistently. They are functions of the process mean and the random process variation. A Six Sigma process that is operating properly makes two nonconformances per billion opportunities. Under Motorola's worst-case assumption of a 1.5 sigma process shift it makes 3.4 defects per million opportunities (DPMO).

11. These are:

$$C_p = \frac{USL - LSL}{6\sigma}$$

$$CPU = \frac{USL - \mu}{3\sigma} \qquad CPL = \frac{\mu - LSL}{3\sigma} \qquad C_{pk} = \text{Min(CPL, CPU)}$$

CPU and CPL reflect the process' ability to meet the upper and lower specification limits respectively. These calculations rely on the assumption that the process data follow a normal (Gaussian, bell curve) distribution. Levinson (2000a) discusses what to do when the data follow the gamma or Weibull distribution.

12. Rearrangement applies to equipment that is easy to move. Fixed-position "monuments," and also utility-intensive semiconductor processing workstations, are not amenable to this. If equipment is mounted on reinforced concrete supports, requires a crane to move it, or has a lot of plumbing for chemicals as well as water, layout changes may not be practical.

13. R. E. Flanders, not Walter E. Flanders of Ford Motor Company. Per Sorensen (1956), Walter Flanders did some work with machine layouts to reduce material transportation distances. Standard and Davis (1999, 16) credit Walter Flanders with arranging production equipment in the order of operations.

14. The z-chart is a control chart for a tool that processes products that have different specifications. For a subgroup of n measurements on product k, the standard normal deviate

$$z = \frac{\bar{x} - \mu_k}{\dfrac{\sigma_k}{n}}.$$

μ_k and σ_k are the mean and standard deviation for product k. The control limits are simply ± 3.

15. The route sheet lists all operations (including value-adding ones like assembly and non-value-adding ones like inspection) that are necessary to make the product. A route sheet that includes specific methods of operation and labor standards is known as a process sheet (Heizer and Render 1991, 277–79).

16. *The Goal* used a Boy Scout hike to illustrate the theory of constraints. Herbie was carrying a lot of equipment and this made him the slowest hiker. Herbie thus limited the rate at which the troop could move. The only way to make the hike (process) go faster was to distribute Herbie's load so he could go faster, that is elevate the constraint.

5

The Theory of Constraints

The theory of constraints (TOC) says that no manufacturing system can work faster than its slowest operation. The slowest workstation is the constraint, or rate-limiting step. Production management through synchronous flow manufacturing (SFM) allows the constraint to set the pace for the entire process. SFM is a "pull" production system like kanban because the constraint pulls work into the production line. That is, the constraint sets the pace for the entire process. To understand SFM, one must first understand the theory of constraints. Figure 5.1 is a manufacturing process that goes through three workstations.

The product requires

- A: 2 minutes
- B: 4 minutes
- C: 3 minutes

Station B can produce only 15 pieces per hour, which means the process can make only 15 pieces per hour. Station B is the *constraint*, or rate-limiting operation.

If the factory runs station A at full capacity to achieve "full equipment utilization," a pile of inventory will grow rapidly in front of station B. Station C, meanwhile, can process 20 units per hour but B can send it only 15. C cannot, therefore, achieve better than 75 percent efficiency.

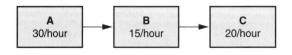

Figure 5.1 Constrained manufacturing process.

This sounds straightforward but the performance measurements of many organizations would actually encourage the workers at station A to run the tool at full capacity.

The next section will compare TOC and its production control system, synchronous flow manufacturing, with JIT. SFM recognizes that the production system cannot run more quickly than the constraint, and it lets that single operation set the pace for the entire process. JIT has each workstation set the pace for the preceding one through kanban and similar controls.

TOC AND JIT

TOC/SFM and JIT are very similar and compatible (Schragenheim and Dettmer 2001). They are both pull systems that seek to reduce inventory. There are some semantic differences; JIT has traditionally relied on kanban, or work orders from *each* downstream process, while SFM uses a "rope" to tie production starts to only one downstream process, the constraint. In today's era of computerized production management systems this difference becomes unimportant. The basic idea is that downstream processes pull work into the production line, the production control department does not push it in.

Unlike JIT, TOC challenges traditional notions about product costing. The section on the cost model shows that cost accounting data can be very misleading when it actually comes to running a production line. Cost accounting does not treat marginal costs and revenues (the differential cost and revenue from making and selling each additional unit). It does not recognize the opportunity cost of missing the chance to sell a unit. It can drive dysfunctional behavior like making inventory to absorb or distribute overhead costs.

Schragenheim and Dettmer (2001) point out that JIT does not recognize the need to protect the constraint with an inventory buffer. A key canon of TOC is that time lost at the constraint is lost forever, and this chapter introduces the related concept of *opportunity costs*. JIT and TOC agree, however, that unused capacity at non-constraint operations is not a problem. One should not run equipment simply to keep it busy, because doing so generates piles of inventory.

TOC says that balanced workstation capacities, a common goal of many factories, are rarely achievable in practice because of random fluctuations in production rates. Instead, efforts should focus on increasing the constraint's capacity, or "elevating the constraint." This is the only way to increase the factory's overall capacity.

Schragenheim and Dettmer (2001) conclude, "The similarities between JIT and TOC are many. The areas of difference are few. Key differences lie primarily in considerations that TOC addresses and JIT does not. The underlying reason for these differences is that by its very nature, TOC takes a broader view of systems management than does JIT."

The Constraint Buffer Provides an Advantage over JIT

Goldratt and Fox (1986, 96) say that both Henry Ford and Taiichi Ohno used a rope system—the connection of consecutive operations through timing or kanban—to run a low-inventory process. Ford's assembly line ran like a clock, with each operation's timing set to generate the right number of parts per hour. It was, at least in theory, a balanced system, and it apparently worked well enough to allow Ford to proclaim that he did not own a single warehouse. Ford's books do not say what happened if an operation went down, although continuous cleaning and maintenance may have prevented considerable downtime. In the Toyota production system, *each* operation controls the rate of the one upstream. The upstream operation does not make anything until the downstream one empties a kanban square (an "in box" for the downstream workstation—the number of kanban squares limits the work that can accumulate) or sends a kanban card to request more work. Ohno mentions "stopping the line," which suggests that the entire line must stop if something goes wrong at one operation.[1]

Synchronous flow manufacturing, which puts into practice the implications of TOC by letting the constraint set the production line's pace, offers two advantages over the Ford and Ohno systems:

1. *Production does not cease if a stoppage occurs anywhere but at the constraint.* The constraint can work from the inventory in its buffer and, since the constraint is the slowest operation, upstream or downstream stoppages don't matter. (Excessive downtime at an upstream operation will, of course, shut down the constraint because the constraint will eventually exhaust its buffer. The buffer size should account for all reasonable upstream contingencies—and shorter upstream cycle times may allow a smaller buffer.) When the stoppage is corrected the non-constraint operation can use its excess capacity to make up the lost work.

2. *Production control is simpler because there is only one connection, the one between the constraint and the first operation.* There is no need for kanbans or kanban squares between any other pairs of operations.

The constraint buffer concept was, interestingly enough, identified more than eighty years ago:

> In a factory making automobile parts the idle machine hours amounted to thirty-six percent on certain machines, which were the "neck of the bottle" in the production scheme.
>
> Before I investigated conditions here, the owners had planned to increase the number (Basset 1919, 82–83). [The instinct to put in more equipment to increase capacity should be familiar to readers of Goldratt and Cox's *The Goal*.]

An analysis of the idle time showed the following breakdown, and we have added familiar modern techniques for dealing with these items:

- Changing dies, 14 percent (Single-minute exchange of die, SMED)

- Material shortages, 10 percent (The constraint buffer)

- Breakdowns, 7 percent (total productive maintenance, TPM)

- Labor shortages, 3 percent (staggered meal and rest breaks, as described by Goldratt and Cox 1992)

- Miscellaneous, 2 percent

Basset continues by proving that he (1) identified the constraint or bottleneck operation and (2) provided an inventory buffer to avoid starving the constraint:

> . . . The change was very simple. It involved merely a change in view-point—considering the machine in relation to the work, which meant providing dies and material beforehand and seeing that they were ready when needed.

TOC is straightforward and logical; why don't more companies use it? We next explore a possible reason: the negative effects of dysfunctional performance measurements.

PERFORMANCE MEASUREMENTS: THE COST MODEL

1. A cost accounting system is not a suicide pact.

2. Be careful what you wish for, you might get it. (This is especially true of business performance metrics.)

The traditional cost accounting system has one legitimate purpose: to prepare tax returns, financial reports for stockholders, and quarterly and annual reports as required by the Securities and Exchange Commission. Investors who rely on such reports to pick stocks should pay close attention to Henry Ford's observation (1930, 25): the law required valuation of plant and equipment in dollars according to accounting methods that were meaningless for practical decision purposes. Their actual value depends on what we can do with them.

Do not let the cost accounting system run the business. Untold misery results from letting the cost accounting system run the factory. This chapter will show that engineering (or managerial) economics, not cost accounting, must manage the factory.

The theory of constraints (TOC) looks at two models, the cost model and the throughput model. The cost model comes from the cost accounting system. The throughput model begins with the premise that a system makes money through sales. This looks straightforward, and the cost accounting system also likes to see high sales revenues. There are, however, key differences between the models.

The Cost Model and Production Management

Under the cost model, operating expense is the primary consideration. Throughput is second and inventory is third. The management control system looks for department-level improvements in each measurement. "Improve locally" is almost synonymous with *suboptimization*. Executives and senior managers would never say, "Suboptimize the process," but the performance measurements say exactly that.

What's wrong with reducing operating expense? That's how Henry Ford increased wages and lowered prices while expanding his business with retained earnings. Some cost reductions, such as using less material or making less scrap, are genuine. Others, however, take place only on paper. This is the danger of letting the cost accounting system run the factory.

The cost accounting system also treats inventory and work in process as current assets. Inventory is an asset only if it will be eventually sold. The cost accounting system can be extremely dangerous here; it can encourage the factory to make unusable inventory to increase equipment efficiencies and reduce cost-per-piece. The results on paper include high local efficiencies and low unit costs. The practical result is warehouses full of unusable inventory.

Consider the following example:

1. Each employee can make up to 25 pieces per hour

    a. Labor is $15.00/hour

    b. The plant has 6 employees

2. Each finished unit requires $6 of material

3. The cost accounting system divides $1500 of overhead across the day's production

4. The market demand is 1000 pieces per day, at $10 per piece

The cost model says to make 1200 pieces (6 workers × 25/hour × 8 hours). This yields the lowest cost per piece, plus $1570 in inventory. This is not a problem if the inventory can be sold later but remember that the demand is only 1000 units a day. The choice to run at full capacity adds less cash than the decision to make only 1000 units. (See Table 5.1.)

The labor cost is a fixed (or sunk) cost. Although the workers are hourly, the company actually pays them for 8 hours a day. Labor is not really a variable cost even though accounting systems often treat it as such. It is up to management, however, to offer overtime, so overtime is usually a genuine variable cost.

Overhead is a sunk cost. The cost accounting system allocates it to the product but, in practice, it is there no matter what the factory does:

> The foremen and superintendents would only be wasting time were they to keep a check on the costs in their departments. There are certain costs—such as the rate of wages, the overhead, the price of materials, and the like, which they could not in any way control, so they do not bother about them (Ford 1922, 98).

Table 5.1 Results of different production decisions.

Pieces	800	1000	1200
Labor	$720	$720	$720
Material	$4800	$6000	$7200
Overhead	$1500	$1500	$1500
Outlay	$7020	$8220	$9420
Cost per Piece	$8.775	$8.220	**$7.850**
Sales (units)	800	1000	1000
Sales (revenue)	$8000	$10000	$10000
Cash	+$980	**+$1780**	+$580
Inventory	0	0	200 @ $7.85 = $1570

Marginal Costs and Revenues

Marginal (or differential) costs and revenues are extremely important aspects of engineering or managerial economics. Engineering economics is not about cost accounting or making the quarterly report look good, it is *practical economic analysis to guide operational decisions.* It answers questions such as, "Should we make something or shouldn't we? Will a project make money for the company or won't it?" Engineering economics recognizes opportunity costs, or the costs of missed or foregone opportunities, that are invisible to the cost accounting system.

The marginal cost is the cost of making another unit. Marginal revenue is the revenue from selling another unit. This is where the word "differential" comes into play. If n is the production rate (variable) and the plant is currently making N (a specific number, for example, 1000 per day), the following is usually true:[2]

$$\text{Marginal cost} \ = \ \left.\frac{d\text{Cost}}{dn}\right]_{n=N} = \text{material cost}$$

$$\text{Marginal revenue} \ = \ \left.\frac{d\text{Revenue}}{dn}\right]_{n=N} = \text{sales price}$$

$$\text{Marginal profit} \ = \ \left.\frac{d\text{Profit}}{dn}\right]_{n=N} = \text{sales price minus material cost}$$

Suppose that, in the example shown in Table 5.1, another customer offers to buy 200 units a day for $7.00 each. Table 5.2 shows what happens.

Table 5.2 Example of marginal costs and revenues.

	As Seen by Cost Accounting	Initial Situation	Marginal Costs and Revenues
Pieces	1200	1000	+200
Labor	$720	$720	+0
Material	$7200	$6000	+$1200
Overhead	$1500	$1500	+0
Outlay	$9420	$8220	+$1200
Cost per piece	**$7.850**	N/A	N/A
Sales (units)	1200	1000	+200
Sales (revenue)	$11400	$10000	+$1400
Cash	+$1980	+$1780	+$200
Inventory	0	0	0

Since the cost per piece is supposedly $7.85, the traditional cost model would expect an 85-cent per piece loss on the customer's offer. "Sell at a loss, make it up on the volume." The actual result is a profit of $1.00 per unit, or $200. This is because the deal's *marginal* labor cost is zero and the overhead is a sunk cost. Material at $6.00 per unit is the only true variable cost, and it's less than the $7.00 selling price.

Now suppose the customer offers to buy 350 units at this price. Six hours of overtime at $22.50/hour will be required to fill this order. Should the factory offer only 200, or should it take the entire order? Table 5.3 shows what happens.

This is still profitable although the marginal profit on the 150 extra units is only $15.00. Nonetheless, the transaction benefits everyone. The workers make an extra $22.50 each and the company gets $15.00. This is not much money but exclusion of competitors is an intangible benefit of filling this customer's entire order. Why encourage our customer to try 150 units of Brand X? The customer, meanwhile, presumably uses the extra 150 units to satisfy its own customers, pay its workers, and earn a profit for its stockholders. The traditional cost accounting system, meanwhile, is still trying to figure out how this deal makes money while selling below the so-called cost (see Goldratt and Cox 1992, 311–13).

Other Deficiencies of the Cost Model

Goldratt and Fox (1986, 24–25) show how the cost accounting system can block quality improvements. The cost model does recognize the expense of

Table 5.3 Effect of overtime (variable labor cost).

	As Seen by Cost Accounting	Initial Situation	Marginal Costs and Revenues, 200 Units	Marginal Costs and Revenues, 150 More Units
Pieces	1350	1000	+200	+150
Labor	$855	$720	+0	+$135
Material	$8100	$6000	+$1200	+$900
Overhead	$1500	$1500	+0	+0
Outlay	$10455	$8220	+$1200	+$1035
Cost per piece	**$7.740**	N/A	N/A	N/A
Sales (units)	1350	1000	+200	+150
Sales (revenue)	$12450	$10000	+$1400	+$1050
Cash	+$1995	+$1780	+$200	+$15
Inventory	0	0	0	0

scrap. It often includes labor and overhead even though these are not really variable costs. It therefore provides an incentive to reduce scrap. As scrap rates fall, however, incremental improvements become more expensive. It is usually more expensive to go from one to 0.5 percent scrap than from two to one percent. The cost model sets up a break-even point that says, "Go no further." *The problem with finish lines is that competitors sometimes don't see them and they decide to keep going.*

Suppose that each percent of scrap costs $10,000 per year in labor and materials. Also assume that the company uses the payback model for project selection, and that payback must occur within two years.[3] A $20,000 project to reduce the scrap rate from one percent to 0.5 percent saves only $5000 per year. It is not cost-effective under the cost accounting model.

The cost model overlooks the intangible element of customer satisfaction, or lack thereof. 100 percent final testing or inspection may or may not keep defective pieces from reaching the customers. Such testing is an appraisal cost under the cost of quality (COQ) model.

Goldratt and Fox (1986, 26–27) go on to show how the cost accounting system can torpedo efforts to increase inventory turnover. Cost models recognize inventory carrying costs, so there is some incentive to reduce inventory. As with scrap, however, incremental improvements become more costly as inventory levels fall. The factory will eventually hit a break-even point and the cost model will say, "Stop here." Low inventory confers other benefits like faster cycle times, and the cost accounting system does not recognize these benefits.

This section has covered the cost model and its dysfunctional effects. It has also introduced the important concepts of marginal costs, revenues, and profits. The next section introduces the performance measurements of the Theory of Constraints. These differ considerably from the cost model's.

THE THROUGHPUT MODEL

The throughput model also looks at three performance measurements, but the priorities are different. They are, in order,

1. *Throughput, or money from sales.* The measurement is the marginal profit on each transaction, not the cost accounting value. This is the sale price minus the cost of raw materials unless overtime is involved (as shown previously).

Lead time plays a role here because it is the time between an order's placement and its fulfillment (and payment). Goldratt and Fox (1986, 12) mention the idea of negative inventory turns, which means receiving payment for an order before we must pay for the raw materials. Harley-Davidson has actually achieved this.

2. *Inventory*—money that the system must invest in items it intends to sell. "Production lead times and work-in-process inventory are really the same thing. One is a mirror image of the other" Goldratt and Fox (1986, 64). Lead time reduction, for which lean manufacturing provides several techniques, reduces inventory.

3. *Operating expense*—the money that the system expends to convert inventory into throughput. Anything that doesn't help transform inventory into throughput is waste.

Quality Costs and the Theory of Constraints

Rework and scrap are obviously bad but where should the factory focus quality improvement efforts? Scrap is generally worse than rework because scrap is a complete loss. The marginal cost model reveals, however, that rework at the constraint can be worse than scrap before the constraint. Refer to Figure 5.1 again:

The product requires:

- A: 2 minutes

- B: 4 minutes

- C: 3 minutes

Also assume that raw materials cost $5.00 per unit, and processing at each station adds $2.00 per unit in direct costs. Rework at any station costs $2.00 per unit. The product sells for $20.00 per unit. Remember that station B can process only 15 units per hour. Table 5.4 shows the drastic difference

Table 5.4   Cost of rework and scrap in a constrained process (1 hour of work).

	Normal	Station A		Station B		Station C	
		Rework	Scrap	Rework	Scrap	Rework	Scrap
Materials	$75	$75	$75 +$5	$70	$70 +$5	$75	$70 +$5
Processing	$90	$90 +$2	$90 +$2	$84 +$2	$84 +$4	$90 +$2	$84 +$6
Cost	$165	$167	$172	$156	$163	$167	$165
Sale	$300	$300	$300	**$280**	**$280**	$300	**$280**
Profit	$135	$133	$128	**$124**	**$117**	$133	**$115**
Δ	0	($2)	($7)	**($11)**	**($18)**	($2)	**($20)**

between accounting costs and marginal costs. Delta (Δ) refers to the difference in profit due to the quality problem. If there is no scrap or rework anywhere, the profit will be $135 per hour.

- Rework at A is straightforward: Δ is ($2), which is the cost of reworking the piece.

- Scrap at A costs $7, $5 for the wasted raw material and $2 for the direct processing cost.

- Rework at B costs $2—according to the cost accounting system. *The opportunity cost is $11.* This is because the rework uses four minutes of time at B, and the station makes 14 instead of 15 pieces during the hour. *Since B has no excess capacity, the opportunity to make and sell a unit is lost forever.* The $11 opportunity cost is $2 for the rework *plus the $9 the factory could have made on the 15th unit.* This underscores a key lesson from the theory of constraints: "Time lost at the constraint is lost forever." Also, *rework at the constraint can be worse than scrap before the constraint.*

- Scrap at B costs $9 on paper: $5 for wasted raw materials, $4 for the processing at A and B. The marginal cost adds $9 for the lost opportunity to make and sell the 15th unit. If rework at the constraint is bad, scrap there is even worse.

- Rework at C costs $2. C has excess capacity, so the rework's only cost is its direct cost.

- Scrap at C costs $11 under the cost accounting system: $5 for wasted raw materials, $6 for wasted direct processing. The marginal cost adds $9 for the permanent loss of the opportunity to make and sell the 15th unit. Remember that station B *cannot* replace any downstream losses.

Implications for Quality Improvement Efforts

To summarize our two key conclusions:

1. Rework at the constraint can be as bad, or worse, than scrap before the constraint.

2. Scrap at or after the constraint results in opportunity costs. Time lost at the constraint is lost forever. *The cost accounting system does not recognize opportunity costs.*

Therefore:

1. Focus efforts to reduce scrap on the constraint and downstream processes.

2. Focus rework avoidance efforts on the constraint. The payoff might exceed that from scrap reduction before the constraint.

This section has examined the throughput model's performance measurements: throughput, inventory, and operating costs. Application of the previously-discussed marginal cost, revenue, and profit concept shows why rework at the constraint can be worse than scrap prior to the constraint. This suggests that the constraint acquires special importance, and the next section uses TOC and its performance measurements to focus improvement activities.

THE THEORY OF CONSTRAINTS AND IMPROVEMENT PRIORITIES

The previous chapter discussed several quality and productivity improvement programs like SMED, kaizen blitz, and error-proofing. Factories that apply them as stand-alone activities, instead of as synergistic parts of a management system, may find them disappointing. This section shows how these activities and others fit in with the theory of constraints.

The basic idea is that the only way to improve the factory's capacity is to elevate the constraint. Anything that helps the constraint work more quickly, reduces downtime at the constraint, or improves quality at the constraint improves the entire factory's capacity. Make the constraint five percent faster and get the equivalent of five percent more factory without spending a dime elsewhere. Eventually, of course, the constraint will change and improvement activities should then focus on the new constraint.

Improvement activities elsewhere also can be beneficial. SMED at a non-constraint may reduce production cycle times and inventory by facilitating small-lot or single-unit processing, although it will not improve the factory's capacity. Other improvements might make a non-constraint less expensive or reduce its scrap rate. The greatest effect, however, usually comes from improvements at the constraint.

Total Productive Maintenance

Total Productive Maintenance (TPM) seeks to avoid machine downtime and assure high availability (Caravaggio, in Levinson 1998). Preventive maintenance keeps equipment from going down. It also may avoid scrap and rework, since defective or worn-out equipment can damage the work.

Gardner and Nappi (2001) stress the importance of eliminating the root causes of even minor equipment stoppages. These have the following characteristics:

1. They last a few minutes or less.

2. The operator can clear them, for example by overriding or resetting a "nuisance" alarm. This may even become an ingrained or accepted part of the job.

3. No equipment repair or spare part is necessary.

These apparently minor stoppages fit the definition of friction perfectly. They are chronic annoyances that people can work around. They don't get permanent correction because it appears easier to live with the problem than to fix it. They don't have obvious consequences and their effect is similar to that of the boy who cried wolf; people stop taking them seriously.

Minor stoppages are actually the equipment's cry for help: early warnings of more serious problems that can cause scrap, equipment failure, or injuries. They are, at the best, non-value-adding work and they are harmful to morale. Their cumulative effect can be serious, especially at constraint operations as defined by the theory of constraints. For example, 300 2-minute stoppages over six weeks add up to ten lost production hours. National Semiconductor's policy is to identify and eliminate the root causes of even minor equipment stoppages.

Norwood (1931, 10) describes how the Ford Motor Company's River Rouge plant treated stoppages. Workers could stop the line if there was trouble (a level of empowerment that modern Japanese factories made famous), such as materials that were arriving too quickly or anything else that interfered with smooth production flow. Stopping the line lit a warning light in a control booth, and the light showed the stoppage's location. If the stoppage lasted longer than two minutes, it required the attention of a trouble mechanic. If the workers could clear the stoppage more quickly, production resumed. *Either way, however, the stoppage's reason and the amount of lost time were recorded;* people didn't merely work around the problem only to have it keep coming back again and again.

Overall equipment effectiveness (OEE) is a measurement of TPM's effectiveness.

$$\text{OEE} = \text{Availability} \times \text{Operating efficiency} \times \text{Rate efficiency} \times \text{Rate of quality}$$

$$\text{OEE} = \frac{\text{Uptime}}{\text{Total time}} \times \frac{\text{Operating time}}{\text{Uptime}} \times \frac{\text{Total pieces}}{\text{Theoretical throughput}} \times \frac{\text{Good pieces}}{\text{Total pieces}} \Rightarrow$$

$$\text{OEE} = \frac{\text{Operating time}}{\text{Total time}} \times \frac{\text{Good pieces}}{\text{Theoretical throughput}}$$

This chapter has already shown the possibly dysfunctional effects of the operating and rate efficiency measurements. They encourage people to make as many parts as possible even if the parts cannot be used. The rate efficiency measurement also encourages people to wait for full loads. This is a key consideration at the constraint but nowhere else.

Rate of quality should be optimized everywhere because rework and scrap are never good. The previous section showed, however, that these efforts must be focused according to the theory of constraints. Efforts to increase availability must focus on the constraint because *downtime at the constraint is almost equivalent to shutting down the entire factory.*

The System Company (1911a, 44–45) provides a clever example of keeping equipment available. It incorporates the SMED philosophy by externalizing preventive maintenance. A cement factory's rotary kilns relied on fans to deliver steady blasts of pulverized coal. The fans and their motors required frequent maintenance and repair due to the high operating speeds. An unplanned stoppage would waste the kiln's entire charge, and the spoiled material could not even be removed until the kiln had cooled for about 36 hours. The factory addressed this problem by providing a truck with an auxiliary blower that could be connected to any kiln. This achieved two purposes when a kiln's fan machinery needed repairs or maintenance:

1. The batch of cement was saved

2. The kiln could be recharged without loss of time

It was furthermore unnecessary to supply *each* kiln with an extra motor; only one extra motor was necessary. The applicable reliability model is the standby redundant system, in which a backup unit comes online when a primary unit fails (or is taken down for maintenance). In this example, several primary units share one backup unit.

We cannot overemphasize the need to apply the underlying principles of total productive maintenance and not merely the specific examples. Your company probably doesn't use cement kilns but the strategy is applicable to a wide range of equipment. Consider, for example, semiconductor processing equipment that requires vacuum pumps. Use the same thought process behind SMED. Which part of the equipment actually adds the value, the vacuum pump or the process chamber? How expensive is the pump relative to the rest of the tool? Could provision of one extra pump for a group of vacuum tools allow externalization of maintenance and repair, and thus improve availability?

SMED and the Theory of Constraints

The only way that single-minute exchange of die can increase capacity is through application to the constraint:

> From the TOC perspective, there's no reason to concentrate on setup reduction unless the setup time happens to be the factor that truly blocks the organization from doing more. Certainly, reducing the setup at the capacity-constrained resource is much more important than reducing setup time on any resource with considerable excess capacity (a nonconstraint). Expending efforts to reduce setups on nonconstraints might rob time from efforts to identify the company's real blocking factors (Schragenheim and Dettmer 2001).

Chapter 6 on single-unit processing will show, however, that SMED can reduce production lead times by reducing batch sizes. Lead times are not necessarily related to capacity. If work has to wait for a large batch operation (like a heat treatment oven) this slows the progress of individual jobs even if it does not affect the factory's net capacity.

Quality and the Theory of Constraints

When the factory is running at full capacity, it is not enough to keep defective parts from reaching the customers. Scrap or rework in the constraint, and scrap after the constraint, cannot be replaced. Anything less than 100 percent rate of quality in or after the constraint imposes huge opportunity costs. It is therefore not enough to catch the defects; the manufacturing system must prevent them.

We have shown that the only way to increase the factory's capacity is to improve the constraint's capacity, that is by elevating the constraint. Preventive maintenance is helpful anywhere because it reduces unplanned downtime (and consequent accumulation of inventory and cycle time) and it may prevent defects. The constraint should, however, receive priority for PM. Single-minute exchange of die can reduce cycle time and promote smaller lot sizes anywhere but it improves capacity only at the constraint.

APPLICATIONS TO PRODUCT AND PROCESS DEVELOPMENT

The theory of constraints introduces several considerations into product and process development. Goldratt and Fox (1986, 48) show why low inventory reduces product introduction time (and, by implication, product development time). Suppose that a factory contains a lot of inventory and work in process. An engineering change (EC) takes place or product engineers want to build prototype parts to test a new design. Production control starts a lot that contains the EC or the prototypes. One of two things must happen.

1. The lot must be expedited, or given priority so it can move past all the inventory and work in process.

2. The lot will have to wait until all the work ahead of it has been processed. The result is *slower feedback on the EC or new product design*. A key principle of development activities is that the competitor who can perform the most plan–do–check–act (PDCA) cycles per unit time has the advantage. Making prototypes wait in long production cycles concedes this advantage to competitors.

The authors also point out that, if an engineering change improves the product, the company faces the choice of (1) finishing the now-obsolete WIP and shipping it to customers, or (2) scrapping the obsolete inventory. If the inventory and work in process are not there, this is not a consideration.

Shorter cycle times also facilitate process and product development by shortening the feedback loop. Shorter PDCA (or its military counterpart OODA: observation, orientation, decision, action) loops can be a decisive competitive advantage. We have already seen how lean manufacturing techniques can shorten cycle times.

Constraints and Design of Experiments

Design of experiments is an important technique for improving products or processes. Beware of designed experiments that require non-saleable parts to go through the constraint. They have the same effect (opportunity costs) as scrap or rework in the constraint. The best time to run such experiments is during times of low demand for the product; that is, when the market is the constraint.

ENDNOTES

1. Norwood (1931, 10) reports that workers at Ford's River Rouge plant were empowered to stop the line in the event of trouble. Any such stoppage resulted in a report, presumably for follow-up and permanent corrective action of the underlying root cause.

2. If hourly workers must work overtime to make more pieces, the marginal cost must include the differential labor cost. Since management can vary overtime, overtime is a variable labor cost. There also may be other exceptions to the equations but they are applicable in most cases.
3. Payback is a very simplistic model, and it is generally inferior to engineering economic analysis (net present worth). It focuses on risk in terms of the time necessary to recover an investment. If a $20,000 project saves $10,000 per year, payback is two years.

6

Single Unit Processing: One-Piece Flow

Chemical engineers are very familiar with continuous processing. Anything that flows or pours is amenable to true continuous flow manufacturing. The plug flow reactor (PFR) is essentially a pipe that may contain a catalyst. Liquid or gaseous raw materials enter at one end and product comes out the other. The flow is continuous, exactly as if the materials were flowing through a pipe. Process control is relatively simple; automatic equipment can adjust process conditions automatically to optimize product quality.

Chemical factories also process liquids in continuous stirred tank reactor (CSTRs). The fluidized bed reactor is the CSTR's analogue for gas processing. Production is absolutely continuous. Raw materials and final products move along in a continuous stream, like water flowing through a pipe (Figure 6.1). *The PFR and CSTR are usually superior to the batch reactor for large-scale production.* Continuous processing also facilitates the use of automatic process control equipment like proportional, integral, and derivative (PID) controllers. These existed before digital computers were readily available.

Traditional manufacturing industries make discrete products like mechanical subassemblies, printed circuit boards, and packaged semiconductor devices. Statistical process control (SPC) does for discrete manufacturing processes what automatic process control does for continuous ones.

Intuition says that process control becomes easier when a manufacturing process starts to behave like a continuous flow process. This means single-piece processing is better than lot or batch processing. Machines and tools that handle one piece at a time are often better than those that deal with batches of parts. Schonberger (1982, 104–105) cites the desirability of making discrete-unit production as much like continuous processing as possible. He also (p. 127) notes the abundance of automatic quality control and

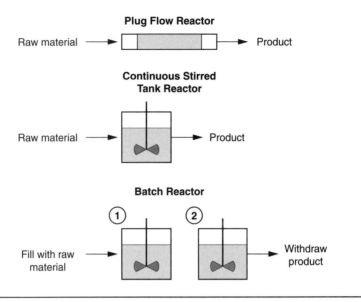

Figure 6.1 Chemical processing options.

monitoring devices in continuous process industries and the desirability of automatic quality checks in discrete-unit production.

Single-piece processing also promotes inventory reduction. "Use of the so-called one-piece flow method dramatically shortens lead times. This is also a major factor in eliminating the need for inventory" (Shingo 1986, 267). Henry Ford's assembly line used one-piece flow. It was, in fact, about the closest to which a manufacturing process could approach true continuous flow:

> We start with the blast furnace and end with a completed motor stacked in a freight car. The casting leaves the foundry on a moving platform or conveyor to one of the assembly lines, it is machined, the other parts are added as it moves along, and when it reaches the end of its line, it is a completed and tested motor— and all of this without a stop (Ford 1926, 108).

Note the words, "all of this without a stop." The parts moved like water through a pipe; they imitated a chemical process as closely as possible. Schonberger (1982, 16) cites exactly the same principle: "The JIT ideal is for all materials to be in active use as elements of work in process, never at rest collecting carrying charges."

Suzaki (1987, 187) states,

If we can tie all production operations with an "invisible conveyor," producing one piece at a time, and transporting one piece at a time as required by the next process, we would have the ideal production system. If we view inventory as waste, then transportation may be viewed as a way to accumulate inventory on the move—which is, of course, undesirable.

This was the purpose of Ford's assembly line. Again note that work in transit is inventory or "float."

Figure 6.2 shows an example of one-piece flow in a semiconductor factory. The equipment spin-coats silicon wafers[1] with photoresists and other coatings; it processes wafers one at a time.

Single-unit processing is a beneficial approximation to continuous-flow processing, which is characteristic of the chemical process industries. Process control often becomes easier as conditions approach continuous flow, and it reaches its ultimate in the controls that chemical factories often use. We will next examine the disadvantages of batch processing, which often increases cycle time and complicates statistical process control.

Figure 6.2 Wafers on a spin-coating track.

Source: Levinson. *Leading the Way to Competitive Excellence*. Milwaukee: ASQ Quality Press, 1998.

THE EVILS OF BATCH PROCESSING

Goldratt and Cox (1992) reveal some of the serious drawbacks to batch processing. Placement of less than a full load in a process tool wastes some capacity. Waiting for a full load wastes time. The theory of constraints demonstrates that only the constraint operation needs to run full loads, with no delays between them. Figures 6.3 through 6.5 show examples of batch processes.

In Figure 6.3, silicon wafers are placed in a vacuum chamber where metal is sputtered or evaporated onto them. The metal becomes the wiring of the semiconductor devices (chips or die). If places on the domes are empty, this capacity goes to waste. Waiting for a full load wastes time. Suppose that each dome holds 12 silicon wafers, the vacuum chamber accepts three domes, and 20 wafers are ready for processing. Waiting four hours for 16 more wafers to show up wastes 80 wafer-hours. Starting immediately wastes 20 units of capacity. Even more complexity results from the fact that the wafers come in lots of varying sizes, and all the wafers in a given lot should ideally be processed in the same batch. The same considerations apply to Figures 6.4 and 6.5.

In Figure 6.4, reactive ionic plasma etches either metal or an insulator from the silicon wafers to define features like wiring and connection positions.

Figure 6.3 Metallization domes and evaporator or sputterer.

Source: Levinson. *Leading the Way to Competitive Excellence.* Milwaukee: ASQ Quality Press, 1998.

Figure 6.4 Wafers and plasma reactor.

Source: Levinson. *Leading the Way to Competitive Excellence.* Milwaukee: ASQ Quality Press, 1998.

Figure 6.5 Wafers in front of a tube furnace.

Source: Levinson. *Leading the Way to Competitive Excellence.* Milwaukee: ASQ Quality Press, 1998.

In Figure 6.5, wafers are heated in a tube furnace. Steam may be passed over the wafers to oxidize the silicon into silicon dioxide. Alternately, heat alone causes dopants (impurities that give the silicon its electrical properties) to diffuse inside the silicon.

Batch Processing and Constraint Efficiency

Suppose that the constraint is a heat-treatment oven that can hold 10 pieces, and the work in Figure 6.6 is waiting for processing there.

Lot 3 cannot be batched with lot 2 to fill the furnace because the lots require different temperature schedules. Lot 1 can be split to run four pieces with lot 2 but there is no way to process lots 1 and 2 in fewer than two runs. Lot 3 must run separately. The oven has a theoretical capacity of 30 in three runs but only 18 pieces are actually run. This makes the constraint's rate efficiency only 60 percent, which is like having the factory shut down 40 percent of the time! Goldratt (1992, 146–47) provides an example of this problem.

The constraint should *always* have a full load. This is achievable by, for example, running 10 pieces from lots 1 and 2 and hoping that six new pieces will show up that require temperature schedule A or B. This requires splitting lots or planning production so there are always 10 pieces in a lot. Nonetheless, the need to run batches obviously complicates production planning for this operation.

Waste of time or capacity at any operation other than the constraint is not really a problem—unless it creates a new constraint. Suppose that a machining operation's capacity is 24 pieces per hour and the heat-treatment oven's is 30 per hour as described previously. The machining center is supposedly the constraint but the heat-treatment step can become the constraint if batch considerations reduce its overall efficiency below 80 percent.

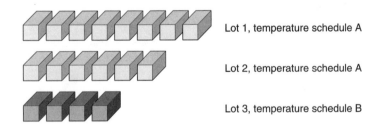

Lot 1, temperature schedule A

Lot 2, temperature schedule A

Lot 3, temperature schedule B

Figure 6.6 Work waiting at the constraint.

If the heat treatment is after the machining step, this won't be a long-term problem because accumulation of work at heat treatment allows the combination and splitting of lots to make up full loads. If it's before the machining step, short-term inefficiencies can starve the machining center (that is, starve the real constraint). This is why the theory of constraints calls for an inventory buffer in front of the constraint. The constraint must operate with full loads and with no waste of time between loads. This is much easier to achieve with single-piece processing.

Batch Processing versus Lean Manufacturing

A single inventory buffer at the constraint can solve net capacity problems that batch operations cause. *It takes only one batch operation, however, to stick an otherwise single-piece-flow system with work-in-process inventory* (Figure 6.7). Schonberger (1986) cites a heat-treatment operation at Omark's Guelph, Ontario facility. Omark knew about lean manufacturing; it was among North America's most advanced JIT plants. It reduced lead time for saw chain orders from 21 to three days in six months but then further progress became difficult.

The problem was a centralized heat-treatment oven that required large batches. "If a large oven heat treated a thousand blades per batch, a thousand would come out at a time." The blades from the otherwise single-piece factory had to accumulate at the oven until there were a thousand. This probably didn't affect overall factory capacity because proper constraint management would have kept downstream operations running. The

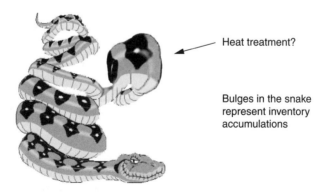

Heat treatment?

Bulges in the snake represent inventory accumulations

Figure 6.7　Even one batch operation can interfere with smooth product flow.

problem was that they were running on three-day-old jobs instead of what could have been same-day jobs.

When the thousand parts came out of the oven they could proceed into the subsequent single-piece operations only one at a time. This situation resulted in waiting time both before and after the heat-treatment operation.

Omark considered several innovative ideas. The first was to install small laser heat-treating stations that could presumably process one unit at a time. The second was to replace the big oven with twenty smaller ones. Omark's purchasing department then found a source of pretempered steel that did not require heat treating. This allowed the company to cut its lead time from three days to one.

Batch Processing and Cycle Times

The connection between cycle time, or the time between an order's production start and its completion, and capacity seems obvious: if there is not enough capacity new orders have to wait or, through expedition, delay orders that are already in the queue. This leads to back orders and delivery delays. One might expect the cycle time to correspond to reciprocal capacity, or time per unit. *The presence of even one batch operation in the process stream can, however, increase cycle times without affecting the capacity at all.*

Return to the example of a metalworking operation that includes a heat-treatment step. For simplicity, Figure 6.8 assumes:

- Machining requires two minutes.

- Heat treatment requires 10 minutes for five parts. The oven is the constraint and it requires full batches to avoid limiting the factory's throughput. This is, of course, the fly in the ointment. We will find that "10 minutes per five parts" equals "two minutes per part" in terms of the plant's capacity. The two are quite different, though, when it comes to lead times.

- Assembly requires one minute per part.

The factory should produce one part every five minutes and it does, but this does not mean it takes a part five minutes to get through the line. Examination of the process times show that it "should" take a part 13 minutes to get through the line: two for machining, 10 for heat treatment, and one for assembly. The factory is progressive and it's quite willing to start single-piece jobs. Consider the progress of five individual jobs through the production line. (See Figure 6.8.)

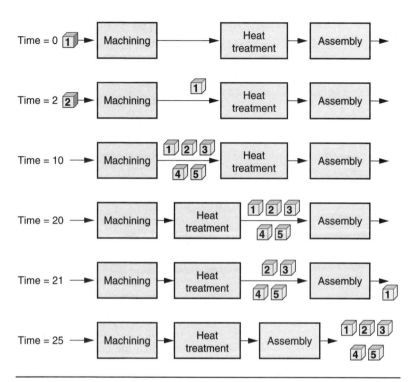

Figure 6.8   Effect of a batch process in single-piece flow.

On the surface, there's nothing wrong here. It took 25 minutes to process the five pieces. There's a big difference, however, between the statements, "The factory makes one piece every five minutes" (true) and "it takes five minutes for a piece to get through the factory" (false). Here's what actually happened. Job 1 started at $t = 0$ and finished at $t = 21$; it took not five but 21 minutes to get through the process. Job 2 began at $t = 2$ and finished at $t = 22$, so it needed 20 minutes. Job 5 started at $t = 8$ and finished at $t = 25$, for a total of 17 minutes. The average time was 19 minutes (also the time for the third job), or 6 more than the expected 13 minutes.

The need to batch-and-queue for heat treatment penalizes us in the preceding and following operations. The oven can't run until it has five pieces. Job 1 is ready for heat treatment at $t = 2$ but it has to wait eight minutes for the other four. Job 5 doesn't have to wait at all. On average, the jobs waited $(8 + 6 + 4 + 2 + 0) \div 5 = 4$ minutes for heat treatment. When they emerged from the oven they waited an average of $(4 + 3 + 2 + 1 + 0) \div 5 = 2$ minutes for assembly. This is why they took an average of six minutes longer than expected.

Replacement of the oven by an induction hardener or similar single-piece tool that requires two minutes per piece—the same nominal capacity as the oven—would actually move each piece through the line in five minutes. The factory's capacity would not change but there would be less inventory and much shorter cycle times.

Batch Processing and Statistical Process Control

Batch processing complicates SPC and capability index calculations. It may even preclude reliable capability index calculations, and this can rule out claims of a "Six Sigma" process. (The remainder of this section may be omitted without loss of continuity if the reader is not familiar with statistical process control.)

Let x be the measurement or dimension of interest. Traditional SPC assumes $x \sim N(\mu, \sigma^2)$ which is shorthand for, "x follows a normal, or bell curve, distribution with mean μ and variance σ^2." (The variance is the square of the standard deviation.)

The control limits for SPC charts are functions of the mean and variance. The capability indices, which reflect the process' ability to meet specifications, depend on the mean, variance, and specification limits. These calculations are very routine when the process follows a normal distribution.

C_p is a ratio of the specification width to the process variation. CPL and CPU measure the process' ability to meet the lower and upper specifications respectively. C_{pk} is the minimum of CPL and CPU. In a Six Sigma process, the variation is so small that, as long as the process is in control, no more than one unit per billion will exceed either specification limit.[2]

In Figures 6.3 and 6.4, however, there are *nested variation sources*. There is variation within batches and between batches. Mathematically,

$$\mu_{batch} \sim N\left(\mu_{process}, \sigma^2_{between\_batch}\right)$$

This is the batch's mean

$$x \sim N\left(\mu_{batch}, \sigma^2_{within\_batch}\right) \Rightarrow x \sim N\left(\mu_{process}, \sigma^2_{within\_batch} + \sigma^2_{between\_batch}\right)$$

For a sample of n measurements from a batch,

$$\bar{x} \sim N\left(\mu_{process}, \frac{\sigma^2_{within\_batch}}{n} + \sigma^2_{between\_batch}\right)$$

For capability index calculations, the correct variance is

$$\sigma^2_{within\_batch} + \sigma^2_{between\_batch}.$$

For example,

$$C_p = \frac{USL - LSL}{6\sqrt{\sigma^2_{within\_batch} + \sigma^2_{between\_batch}}} \text{ and}$$

$$CPU = \frac{USL - \mu_{process}}{3\sqrt{\sigma^2_{within\_batch} + \sigma^2_{between\_batch}}}$$

where USL and LSL are the upper and lower specification limits respectively.

Fairchild has extensive experience in dealing with nested variation sources. The StatGraphics software package (as well as others) can sort out the within- and between-batch variation. Very few books, however, deal with this situation and it often results in confusion.[3] There are often too few batches to get a good estimate of the between-batch variation. This is usually the larger of the two variance components so the control limit and capability index calculations must rely on a very nebulous estimate.

Fairchild Semiconductor's Mountaintop plant also has developed and applied SPC and capability index methods for nonnormal distributions.[4] It is not yet known, however, what to do with nested nonnormal distributions, or nested distributions in which the batch averages follow the normal distribution and units within the batch follow, for example, a Weibull distribution.

Finally, batch processes complicate interpretation of the chart for variation (sample standard deviation or range). The sample variation for each batch reflects only the batch uniformity, *not the overall process variation.* Consider, for example, a heat-treatment oven the temperature of which varies no more than one degree across the batch. The batch uniformity is very good. Suppose, however, that a problem with the temperature controller causes the batch-to-batch temperature variation to double. A variation chart that treats individual pieces from one batch as an independent sample will never detect this change if the within-batch uniformity stays the same.

The tube furnace in Figure 6.5 is worse than a batch process. If a gas like steam is flowing through the tube to oxidize silicon, its composition changes between the entrance and the exit. There is systematic variation from one end of the tube to the other. This leads to a *multivariate normal distribution* (Figure 6.9).[5] Advanced statistical process control methods, such as Hotelling's T^2 chart, exist for these applications. Process capability

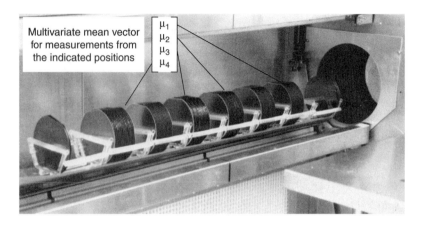

Multivariate mean vector for measurements from the indicated positions

$\begin{matrix} \mu_1 \\ \mu_2 \\ \mu_3 \\ \mu_4 \end{matrix}$

Figure 6.9 Multivariate distribution for a batch process.

Source: Levinson. *Leading the Way to Competitive Excellence.* Milwaukee: ASQ Quality Press, 1998.

index calculations for such processes are, however, still in the realm of applied research journals.

This section has shown how batch-and-queue operations can add cycle time, complicate statistical process control, and even add an extra variation source (between-batch) to process variation. There are many reasons for switching to single-unit or at least small-lot flow if it is feasible. The next section introduces kanban and just-in-time production control. These help avoid inventory accumulation, although they cannot overcome completely the harmful effects of batch operations.

KANBAN AND JUST-IN-TIME

Kanban, JIT, and synchronous flow manufacturing (SFM, the subject of the next chapter) are *pull* systems. A *pull system* requests raw material when it is hungry, that is when workstations have excess capacity. In practice, the constraint operation should never wait for work. There should always be an inventory buffer at the constraint but nowhere else. Kanban, JIT, and SFM are all pull systems. A push system does exactly that; it pushes raw materials into the factory in response to sales forecasts or orders. Constrictor snakes that swallow their meals whole are designed to work this way (Figure 6.10).

Other systems can suffer serious indigestion from this approach. Goldratt and Cox (1992) show manufacturing management expediting certain orders to get them out in time while other orders fall even further behind. The bulge in the snake is a big inventory bubble that must make its way through the system.

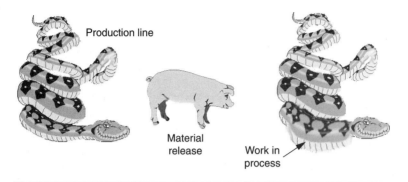

Production line

Material
release

Work in
process

Figure 6.10   "Push" manufacturing system.

It has a long cycle time; constrictor snakes often sleep for weeks or even months during the digestive process. This is not something a factory should do. JIT systems, and especially single-unit processing, assure a steady flow of materials through the line. They leave "pig-swallowing" to systems that are designed to handle it.

Kanban

Kanban means *card*. The card is the production system's order for more raw materials or parts. An empty or, to use the snake analogy, hungry, workstation sends a card up the line to ask for more parts. A computerized production management system can take the place of a physical card. Another method is to put a flag on top of the last one or two pallets or boxes of inventory. As long as boxes or pallets cover the flag, upstream stations do not make more inventory. Exposure of the flag means the workstation will soon run out of work. When this happens, the workers put the flag on a post to signal upstream operations to make more pieces (Heizer and Render 1991, 572–73).

Ohno (1988, 40–41) provides the following rules for kanban:

1. A kanban is a withdrawal order, delivery order, and work order.

2. No one can make a part without a kanban for it. (Don't make anything that isn't needed.)

3. A kanban must be attached to each part or lot.

4. Everything that is produced in response to a kanban must be defect-free.

5. Reduction of the number of kanban promotes operational improvement.

Kanban uses small batches, often only a few hours' worth of production. Since the batches or lots are small, machine setups are frequent. Frequent setups are a key objection to small-lot or single-unit processing. Goldratt's and Cox's *The Goal* (1992) begins with an argument over a tool setup; the factory must choose between delaying an emergency job or losing a previous setup for a less important job. SMED promotes JIT by reducing setup times.

There is little difference between kanban and SFM. In SFM, the constraint is the only operation that can request more production starts. Inventory is kept only in front of the constraint, which must never run out of work.

Pull Production Control in a Blacksmith Shop

Here is an early 20th-century example of pull production control:

> A large blacksmith shop has cleverly solved the problem of keeping the workmen supplied with jobs without loss of time to the pieceworker.
>
> Near each shear or cut-off in the shop is a push button connected with an indicator in the department office. A few minutes before the shear man is through with a job, he pushes a button, which indicates to the storekeeper that shear No. 12, for example, will soon be ready for stock for another job.
>
> . . . The storekeeper then pushes a second button, which calls the man who has charge of the iron house. To him is given the original memo order, and he delivers the material to shear No. 12, in ample time to keep the workman busy, and prevent any loss of time on the shear (The System Company 1911a, 27).

There was an inventory of iron bars in the stockroom but the system apparently called for their delivery to the workstation on a just-in-time basis. The system was very similar to kanban because the worker called for more work as he finished the work he had, although he used an electronic signal instead of a kanban card.

Small Lots Promote Quality and Productivity

Large lots and inventory conceal problems. Heizer and Render (1991, 571) use the analogy of a stream's water level, with depth representing inventory. Rocks symbolize quality and productivity problems that dam up the stream and make the water deeper. Inventory reduction exposes the tops of the rocks and allows their removal. The stream becomes shallower to reveal a new layer of rocks. The continuous improvement process continues, ideally, until all the rocks are gone.

A key principle of feedback process control is that feedback loops should be short. Delay in information feedback must be avoided. Inventory increases feedback time because it takes longer for work to reach downstream operations where problems might be discovered. Self-inspection at each operation is the ideal way to find problems but, depending on the product's nature, this might not be possible. In semiconductor device manufacture, for example, many electrical characteristics cannot be tested until the product is complete. Western Electric (1956, 217) emphasizes the need for rapid feedback, which shorter lead times support:

1. Delay may mean that unnecessary amounts of undesirable product are produced. This will reduce yields, increase scrap and increase the amount of product rejected by Inspection.

2. Delay may make the cause of trouble harder to identify, and in many cases it cannot be identified at all. It is an axiom of quality control that one should identify assignable causes while those causes are still active.

Suzaki (1987, 212–15) shows that small lots also improve communications. Inventory decouples workstations and operators from each other. People take work from the incoming inventory pile. When it's finished, they put it in the outgoing inventory pile. This reduces interaction and communication. When workers make product only as required by the next operation (their customer), communication improves and quality problems are discovered much more quickly.

The section on 5S-CANDO showed that workplace cleanliness is important because problems like fluid leaks can hide in messy environments. Quality and productivity problems can similarly hide in big piles of inventory but not in a lean manufacturing environment. Small lots and single-piece lots take away the hiding places. It has already been shown that single-unit production makes statistical process control much easier.

Inventory Reduction and Semiconductor Product Quality

Many semiconductor fabrication processes pose challenges because the equipment is designed to handle batches. There are some single-wafer tools but, as was shown earlier, the presence of any batch tools in the process stream creates batches and inventory. Fairchild's Mountaintop plant has nonetheless reduced inventory by using SFM. Work is not started until the constraint operation calls for it.

It is very important to avoid having semiconductors wait once processing has started. Schonberger (1986, 111) points out that this results in more opportunities for exposure to particles, chemicals, and static electricity. The cleanroom reduces the incidence of particles that can damage the product but it cannot eliminate them completely. The less time the wafers spend in a vulnerable condition—between removal from their protective package and completion of the final insulation layer, which seals them against dust and contaminants—the less chance there is that something will happen to them. The same principle applies to anything that can suffer damage or deterioration in storage, especially perishable goods.

The chemical process industry's experience has shown the controllability of continuous-flow processes, and this chapter has shown the benefits of having discrete-unit manufacturing at least approximate this behavior. Batching increases cycle time and inventory, and it complicates statistical process control. The next chapter will describe synchronous flow manufacturing (SFM). This is the well-tested production method that Fairchild's Mountaintop plant uses to manage production.

ENDNOTES

1. The silicon wafers are cut into chips or die at the end of the manufacturing process.
2. "In control" means that the mean and variance are stable. A process is out of control when its mean shifts or when its variation changes. In the latter case, the change is usually undesirable, that is, an increase. If a Six Sigma process suffers a shift of 1.5 standard deviations in its mean, it will make 3.4 ppm nonconforming product.
3. Levinson and Tumbelty (1997) *SPC Essentials and Productivity Improvement: A Manufacturing Approach*, ASQ Quality Press, addresses this issue.
4. If the measurements follow a univariate gamma or Weibull distribution, the control chart's center line is the median (50th percentile) of the data. The upper control limit is the 0.99865 quantile, the same quantile as the mean-plus-three-sigma limit of the Shewhart control chart. The lower limit is similarly the 0.00135 quantile. Capability indices can be deduced from the portion of the distribution that is outside the specification limit(s). See Levinson (2000a).
5. Experience with data from this process suggests that it is indeed multivariate normal. The parameters are the mean vector, or a vector of position-dependent means, and a covariance matrix.

7

Synchronous Flow Manufacturing

Synchronous flow manufacturing (SFM) has delivered proven results at Fairchild Semiconductor's plant in Mountaintop, Pennsylvania. Six months after implementation of the theory of constraints approach to production in 1991, inventory was down 42 percent and throughput was up 28 percent. A change in the production control method, not additional capital spending, therefore increased the plant's effective capacity by more than a quarter. The plant, which had been losing money in what was then Harris Corporation's semiconductor sector, became one of the corporation's revenue leaders (Murphy and Saxena, 1997). This chapter will include not only the theory of SFM but also a section on how the Mountaintop plant uses it.

WHAT IS SFM?

Dr. Eliyahu Goldratt defines SFM as "any systematic way that attempts to move material quickly and smoothly through the various resources of the plant in concert with market demand." SFM designs protective inventory around the *capacity constraint resources* (CCRs), or operations that affect the entire process' throughput significantly and where problems can adversely affect due date performance. Buffers at strategic operations or CCRs clash with traditional thinking on the production line. Instead of stockpiling inventory at all operations where Murphy ("anything that can go wrong will go wrong") can strike, SFM focuses buffers only at operations that can affect overall throughput to the system.

Synchronization: Some Background

Synchronization implies timing and coordination of activities. As a young boy, Henry Ford took watches apart (and put them back together so they worked). "Henry had learned an important lesson. Time was a machine that could be taken apart and put back together" (Gourley 1997, 12). Gourley (1997, 30) continues:

> Inside a watch, the mainspring slowly unwound and drove the clock wheels. The hour wheel made one revolution every hour. The minute wheel moved faster, making sixty revolutions every hour. Henry realized that his factory could operate the same way.
>
> . . . The speed of the work was carefully timed so that the assembly line did not run too fast or too slow. Where the workers put together the chassis, the line moved 6 feet (2 meters) per minute. Where the workers bolted the front axle to the chassis, the line moved faster, 15 feet (4.5 meters) per minute.
>
> It was like setting the mechanism of a clock.
>
> Henry had created a giant moving timepiece.

Suzaki (1987, 136) writes,

> As *rhythm* in music drives the whole orchestra to synchronize with the swing of the conductor's baton, cycle time in a factory will drive the production line with a smooth, steady flow of goods. In fact, the Japanese also refer to cycle time as "tact" time [also takt time—the word is European in origin], reflecting the rhythmical swing of the baton.

The loading and firing drill from von Steuben's (1779) *Regulations for the Order and Discipline of the Troops of the United States* stressed the need for each soldier to count a second of time between each motion. Volley fire would not work if everyone did not work at exactly the same speed. Drums, meanwhile, kept ranks moving in steady lines. Without such synchronization, disorder would occur as faster soldiers got ahead while slower ones fell behind. This theme recurs in the Boy Scout hike in Goldratt and Cox's *The Goal*, where the faster hikers got far ahead of the group while the heavily-laden Herbie (the constraint) retarded the progress of those behind him. The concept of takt time is similar.

The connection between clocks, timing, and Goldratt's and Cox's (1992) drum-buffer-rope production management system is obvious. Synchronization means that the parts of the system must keep pace with each other to avoid jamming (inventory pileups) and stoppages from lack of work.

Three Systems: Ford, Kanban, and SFM

It is worthwhile to compare the original Ford production control system to the Toyota system that it inspired, and also to SFM. The Toyota system relies on pull from downstream workstations, and kanban or the equivalent provide the signal to send more work. SFM relies on pull from only the constraint operation, to which an information "rope" connects production starts.

Henry Ford's assembly line was not a pull system but it was not exactly a push system either. It was the very definition of a balanced factory; all operations had supposedly equal capacities. Operations were timed to give workers exactly the amount of time they needed, no more and no less. If, for example, a job improvement reduced the time for a given task from 200 to 100 seconds, the assembly line's speed was apparently doubled in that operation. If this happened to be the first operation, of course, its throughput remained constant instead of doubling and flooding downstream operations with excess production.

It is very worthwhile to look at the Ford system because, according to the matchstick-and-dice simulation in Goldratt's and Cox's (1992) *The Goal*, it should not have worked. It is very instructive to learn why it *did* work despite the theoretical obstacles. Goldratt and Cox show that balanced operations do not work as expected. The effect of random variations in production speed at individual operations generates inventory and retards production. If each workstation can theoretically process an average of 100 pieces an hour, the factory rarely achieves this production rate. A random slowdown at one station stops production at the next one (unless there is a pile of inventory waiting there), and the system never makes up the shortfall.

Ford (1922, 1926, and 1930) never discussed what happened if a worker fell behind for a moment, a tool broke, or a machine had to stop for maintenance. Did inventory pile up before that operation and continue to accumulate as preceding operations kept working? *The Goal* suggests that this system should not have worked but it apparently worked very well. The reasons were probably as follows, and readers can apply these lessons to their own situations:

1. Equipment received continual preventive maintenance. "Minor" equipment stoppages were apparently never considered too minor for aggressive correction of the underlying causes. As shown in the section on total productive maintenance in chapter 5, Gardner and Nappi of National Semiconductor (2001) reintroduce the same principle.

2. Automation and subdivision of tasks probably suppressed variation in processing times, thus mitigating against the "hurry up and wait" effect of unsteady production flow.

Recall that, per equation 4.1 (page 70), variation in processing times and also in work arrival rates are major driving forces for inventory buildup. Reduction of these variation components to zero would theoretically allow almost 100 percent equipment utilization without inventory buildup. The Ford system was designed to run like a clock, and this suggests the importance of the modern takt time concept. Mechanization of tasks removed the human variability element from processing times and doubtlessly helped achievement of near–100 percent utilization.

Recall also that *workers at the River Rouge plant could stop the line if work was coming too quickly or there was a malfunction*—a level of empowerment that modern Japanese factories made famous. The plant recorded the reasons for all such stoppages, and this presumably led to permanent correction of the root causes. This looks similar to the Japanese practice of pushing a production line to its limits to expose problems that can hide in excess labor or inventory. Per Schonberger (1982), the Japanese will remove workers from a line or reduce the number of kanbans (and thus the protective work-in-process buffer) to do this. The River Rouge plant may have achieved similar results by accelerating the line until alarms began to ring.

Finally, the Ford assembly line was not a single line of sequential operations but rather a set of parallel activities. An image of the bill of materials (BOM) as tiny streams (raw materials, simple parts) that converge to form larger streams (subassemblies) that finally merge to form a river (Model T cars) is probably the best way to describe what took place. Parallel assembly is generally the best way to make anything (Figure 7.1). This also is the preferred approach in complex chemical synthesis. A major challenge in

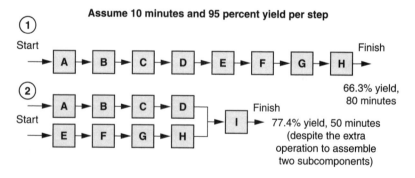

Figure 7.1   Advantage of parallel assembly (or parallel chemical synthesis).

semiconductor manufacturing is that there is currently no way to make the product through parallel operations.

Now consider kanban and SFM. A previous section showed that kanban is a pull system in which downstream workstations demand, or pull, work from upstream ones. Upstream workstations do not do anything until they receive kanban cards or until their operators see empty spaces in the next operations' kanban squares.

In synchronous flow manufacturing, the constraint sets the pace for production starts. Intervening workstations can go at their own pace, they do not need to wait for signals from the next ones in line. Excess inventory cannot accumulate because the constraint limits the production starts. In summary:

1. The Ford system was not a pull system per se. Production starts were, however, at the assembly line's designed pace. The idea of takt time, or time allotted per unit, probably played a role. Preventive maintenance, breakdown of jobs into elemental tasks, and deliberate efforts to eliminate human-induced variation from tasks apparently mitigated against variation in process times and its consequences.

2. Under kanban, downstream workstations pull work directly from their upstream neighbors. Modern Japanese factories often allow a worker to stop the entire line if there is a problem. This does not really reduce output any further because operations that are downstream from the stoppage run out of work. If upstream ones kept going they would simply produce a pile of inventory at the stoppage point.

3. In synchronous flow manufacturing, the constraint controls production starts. SFM offers two advantages over kanban:

   • It is simpler because there is only one link, the one between the constraint and production starts.

   • SFM's inventory buffer allows the constraint operation to keep working despite stoppages at upstream operations. There is no loss of overall throughput, as there might be in a straight kanban system.

Comparison of SFM with the Ford production system is instructive because it reveals potential methods of surmounting obstacles to high equipment utilization. (Remember that high equipment utilization is not a goal in itself except at the constraint operation.)

SYNCHRONOUS FLOW MANUFACTURING: APPLICATION

Refer to Figure 5.1, p. 109. The product requires:

- A: 2 minutes

- B: 4 minutes

- C: 3 minutes

Operation B is therefore the constraint, as it can make only 15 units per hour. Goldratt and Cox (1992) show that there is little benefit in trying to balance or level capacity. The matchsticks-and-dice exercise in *The Goal* shows that balanced systems don't work as expected because of random variations in production speeds. Equation 4.1 (p. 70) shows that waiting times in queue rise rapidly as tool utilization approaches 100 percent. The only way to mitigate this effect is to suppress variation in production speed, and the Ford production system may well have achieved this.

Elevate the System's CCRs through Constraint Management

Once the shift to balancing flow rather than leveling capacity is underway, then the system's constraints can be managed by the following procedural steps, according to Dr. Goldratt:

1. Identify the system's constraint

2. Decide how to exploit the system's constraint

3. Subordinate everything else to the decisions made in step 2

4. Elevate the system's constraints

5. If the capacity of a constraint is elevated to the point that the constraint is broken, go back to step 1. Do not allow the next constraint to remain because of inertia.

Identifying the system's constraints can be a very difficult process. If the wrong constraint is chosen, then it is imperative to select another and continue with the process. Data that identify a constraint may become inaccurate once that constraint is exploited. Inventory levels or "wandering bottlenecks" may direct the attention to another process responsible for the effect that triggered the initial decision.

The section in chapter 9 on linear programming (LP) shows how a well-established operations research tool can identify a factory's constraints. This assumes that accurate data on workstation capacities and each product's workstation time requirements are available. LP also shows how bottlenecks can indeed shift as the product mix changes.

DRUM-BUFFER-ROPE (DBR)

Synchronous flow manufacturing recognizes machine B in Figure 5.1 as the constraint, or rate-limiting operation. Instead of pushing material into the line at A, B pulls material into the line. Machine B beats the drum, or sets the pace, for material starts. If we look at the process as a clock, Machine B is the pendulum that dictates the speed at which all the gears turn.

A *rope* links production starts to machine B. Goldratt and Fox (1986, 88) state that Henry Ford's assembly line used a literal rope, the conveyor belt, to link production resources. Kanban uses logistical 'ropes' to do this.

Idle time at the constraint is irrecoverable and lack of work idles the constraint. An inventory *buffer* waits for processing at B. This is the only planned inventory in the production line and it exists to prevent starvation of the constraint. The inventory buffer decouples the constraint from, or makes it independent of, fluctuations and problems in upstream operations. Unexpected downtime at A, or scrap at A, will not idle the constraint. When the problem at A is corrected, this operation can use its superior capacity to replenish the buffer. This buffer is the *internal constraint buffer*.

Using the DBR System

By identifying the CCRs in the manufacturing process we will be able to handle the inevitable fluctuations and daily disruptions that occur within the plant. According to Dr. Goldratt, the following categories classify these disruptions:

1. *Dependent events.* An event or series of events that must take place before another can begin.

2. *Statistical fluctuations.* Fluctuations that prevent an accurate prediction of critical factors relating to production performance. We have already seen how production speed variations, for example, can negate all assumptions of how a balanced factory should work.

Dependent events and statistical fluctuations are facts of life. The drum-buffer-rope system is not merely a production control system, it is an ongoing improvement process that specifically focuses management decisions to increase throughput and control inventory levels for the organization. The *rope* concept is defined by releasing production starts according to the throughput of the constraint. *Under no circumstances should work be started in the manufacturing process to "keep the workers (and/or equipment) busy."* Remember that work is force times distance, not force alone. Exertion that does not cover distance, in terms of finished goods sales (the throughput metric), is not work. One can even define such exertion as *organizationally-mandated soldiering on a grand scale* (Figure 7.2). (Change agents who are trying to overcome dysfunctional performance measurements like equipment utilization may consider memorizing this phrase.) The people and equipment mark time by making inventory that, at best, has carrying costs, and also wastes materials if it can't be sold later.

In the real world of manufacturing, the system's constraint can be anywhere in the process, and chances are it is not the very first step. The DBR system recognizes this and deals with it as shown in Figure 7.3.

The system's constraint or control point will dictate the introduction of new work into the manufacturing process. All constraints before this operation must be identified and elevated to the point where the manufacturing

Figure 7.2   Different forms of soldiering.

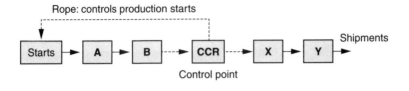

Figure 7.3 Drum-buffer-rope system.

process can support the throughput of the control point. The control point is the system's *drum*. The drum is the rhythmic beat for the system and it dictates the throughput level for the entire manufacturing process.

The *rope* ties the production start operation to the throughput of the control point. The inventory level of the rope is determined by the drum throughput and observed cycle time to that operation. For example, if the drum throughput is 100 per day and the cycle time is 10 days, the rope will contain a 1000 piece inventory.

This underscores a previous observation: an increase in capacity (throughput) does not necessarily reduce inventory or cycle time. SFM and DBR make a good start at inventory reduction, and often cycle time reduction, by limiting production starts to the constraint's throughput. SFM and DBR cannot, however, overcome many of the cycle time problems that come from batch-and-queue operations and long setups. While techniques like SMED increase capacity only when they elevate the constraint, they can reduce cycle time elsewhere and thus reduce inventory.

Inventory levels within the rope are controlled very accurately through the DBR system. *Under no circumstances should work be started that is not dictated by the output of the control point;* let the DBR system do its job. The inventory levels will then take care of themselves.

The *buffer,* or inventory that the factory locates strategically to protect against disruptions in the manufacturing process, ensures the beat of the drum (control point). The buffer can be staged directly at the control point operation or spread out across previous operations. The level of protective capacity is determined by the confidence in the manufacturing process and cycle time days from the start operation to the control point. For example, to create a three-day buffer for a control point set at 100 pieces per day, a 300-piece buffer would be set at prior operations. The location of the inventory would be staged according to observed cycle time data. This brings us back to cycle time. *Shorter pre-CCR cycle times mean smaller acceptable constraint buffers.*

The Shipping Buffer

Since machine C's capacity is greater than B's, it can always make up lost time. A breakdown at C, however, may cause a temporary shortage of finished goods and thus delay shipments. If shipment dates are critical, a *shipping buffer* also may be kept. This is simply an inventory of finished goods. If problems occur downstream of the constraint, this buffer can satisfy delivery requirements for a while. DBR therefore requires inventory in two places at most: the internal constraint buffer and the shipping (finished goods) buffer.

As with the constraint buffer, the shipping buffer also can be spread across non-constraint operations. This is especially true if the operations that are downstream from the constraint are highly reliable. Work that has passed the constraint should move quickly if there are no downstream stoppages. Fairchild Semiconductor's Mountaintop plant spreads its shipping buffer across their downstream operations.

Buffer Management

Buffer management is the selection of appropriate buffer sizes, and management of the entire production pipeline. Goldratt assigns three levels of urgency for buffers: *OK, watch-and-plan,* and *act* (Figure 7.4). The internal constraint buffer and the shipping buffer work the same way.

Work flows from right to left in Figure 7.4, while time increases from left to right. (In the alternative format, work moves from left to right and time increases from right to left.) If the OK zone is not full, this is acceptable. Normal production variation may account for this. Shortfalls in the watch-and-plan zone are a warning. A response plan must be prepared in case the shortage enters the act zone. A shortage in the act zone threatens the constraint's production or the shipment schedule.

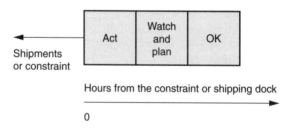

Figure 7.4 Buffer status.

Goldratt and Fox (1986, 122–27) explain the basis of the three zones. The work that is planned for the act zone, the one closest to the constraint, should always be present. This inventory protects the constraint from starvation. It is acceptable and even desirable to have some work missing from the OK zone, the one farthest from the constraint. If the buffer is usually full, the factory is carrying too much inventory as insurance against disruption at the constraint.

Acceleration of production in upstream operations can correct a shortage in the internal constraint buffer. Work also can be expedited through downstream operations, but lost time at the constraint itself can never be made up.

Production Scheduling

Remember that the constraint pulls material into the line: production control does not push it in. Customer orders define the product mix. When jobs enter the production line, the production control system assigns them dates due at the constraint according to historical cycle time data (Rerick and Klusewitz 1996). The date due at the constraint is important because if too many jobs are late the constraint will run out of work. Jobs also may have to wait at the constraint if nowhere else, so those with earlier delivery due dates should take precedence over jobs with later due dates.

Constraint Buffer Graph

The *constraint buffer graph* displays the number of pieces at the constraint and the number of pieces *n* days from the constraint. The graph also identifies parts that are on schedule, early, and late.

Lots are also grouped into three zones according to their relationship to the constraint. The zones prioritize expedition of jobs that are behind schedule. The *expedite* zone encompasses lots in the third of the tracking period closest to the constraint. Late lots in the expedite zone can create holes in the constraint buffer inventory and must be expedited. The *watch-and-plan* zone encompasses in the middle third from the constraint. Late jobs in this zone could create trouble if delays occur at future operations. The *OK* zone is the third farthest from the constraint. Late jobs here can often make up lost time, and it is rarely necessary to expedite them this early in the process. Production personnel should ideally worry only about late jobs in the expedite zone (Rerick and Klusewitz, 1996).

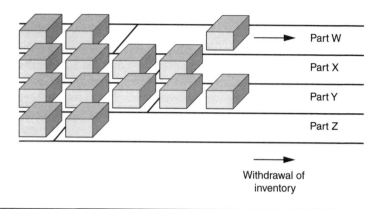

Figure 7.5   Buffer management and kanban.

Buffer Management and Kanban

This buffer management scheme is really not that different from kanban. Heizer and Render (1991, 572) show two inventory management schemes.

The first system depicted in Figure 7.5 does not use an actual kanban to order more material. Boxes of each part are stored in lanes on the floor. A line in each lane marks the reorder point. The line's position depends on the part's consumption rate. Withdrawal of a box of Part W exposes the line and initiates an order for more. The same will happen when the next box of X is taken.

The second system ("Outbound stockpoint with warning signal marker") stacks the boxes of parts on top of each other. Figure 7.5 essentially rotates 90° counterclockwise. A flag is placed at the reorder point, that is, on top of the second box of part W and the third box of part X, under the requirements of Figure 7.5. Exposure of the flag means more parts should be produced.

DBR CONSTRAINT
MANAGEMENT: APPLICATION

The Drum-Buffer-Rope (DBR) Constraint Management Tool provides an environment for scheduling the manufacturing shop floor based on two points in the manufacturing flow. The tool is based on the theory of constraints (TOC) concepts put forth by Eliyahu Goldratt of the Avrham Y. Goldratt Institute.

The DBR Constraint Management Tool has two functions:

1. It assures throughput by managing the constraint buffer to assure that the constraint never runs out of work.

2. It prioritizes individual jobs to help assure on-time delivery to customers. Operators can see any job's due date by selecting the lot detail from a computer screen.

We have seen that a key premise of TOC is that a manufacturing flow will have a single constraint point that limits the output of the entire manufacturing operation. Thus, the management and exploitation of this constraint point provides immediate benefits to the whole system. The means for managing the constraint point is to divide the manufacturing flow into two buffers. We have seen that the constraint buffer protects the constraint from starvation that can result from stoppages in upstream operations. The second buffer is called the *shipping buffer*. The constraint buffer consists of all manufacturing operations from the start of the manufacturing flow, up to and including the constraint operation. The shipping buffer consists of all operations after the constraint to the end of the manufacturing flow. By managing the product inventory levels in the two buffers, the product output through the constraint point can be optimized, resulting in a smooth manufacturing operation that meets business profit and product delivery goals. The DBR Constraint Management Tool is a tool for helping the manufacturing shop floor manage its product inventory and product movement to optimize the throughput of the constraint point and its entire manufacturing operation.

The DBR Constraint Management Tool divides the manufacturing flow into two buffers separated by the constraint point. The tool then applies priority dispatch scheduling algorithms to all manufacturing lots in the flow. Priority dispatch is applied to the end point in the flow as well as the constraint point in the flow. This is different from traditional scheduling in which the priority dispatch algorithm is normally applied to the end point of the manufacturing flow. The net result is that manufacturing lots receive two priority dispatch rankings. One measures the lot's on-time performance to the end of the manufacturing flow (the traditional point). The second priority dispatch measures the lot's on-time performance to the constraint point of the manufacturing flow. The result is better management of lot arrival to the constraint point, ensuring that the constraint is always active. This avoids the previously-mentioned opportunity costs of an idle constraint. By keeping the constraint active, the entire manufacturing operation benefits through higher product throughput and better management of lot due date performance.

The DBR Constraint Management Tool uses a Graphical User Interface (GUI) implemented with PowerBuilder to display the constraint and shipping buffers. The tool provides a graphical view of work-in-process (WIP) in relation to the constraint point. Work-in-process data is fed from Workstream and stored in Oracle database tables. The tool indicates product on-time performance through color-coded bars indicating which product is on time, early, or late. The tool allows the user to quickly locate product lots in the manufacturing flow. Lot priorities are based on the product's status in relationship to the constraint. The pictorial representation of the WIP, the dual classification of product on-time performance, and the ability to quickly identify lots for prioritization make the DBR scheduling tool a unique and valuable shop floor scheduling tool.

The DBR Constraint Management Tool display has two stacked bar graphs that represent the Constraint Buffer and Shipping Buffer (Figures 7.6 and 7.7). The bar graphs quantify the total number of units located in the job flow at various days from the known factory constraint. Time from the constraint or end of the flow is determined by a products-predicted cycle time at the operations in the manufacturing flow.

The time frame before the constraint resource makes up the Constraint Buffer. The Constraint Buffer graph (Figure 7.6) displays the quantity of product and the time in days from the constraint resource. Time from the

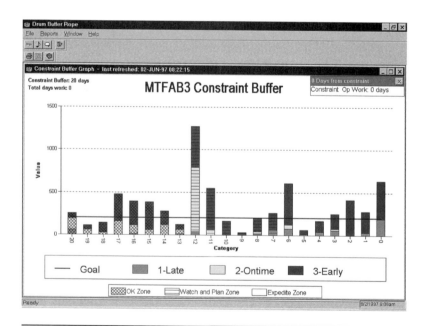

Figure 7.6   DBR constraint buffer.

constraint runs from 20 days at the left to 0 days at the right. Days 20 to 13 are the OK zone, days 12 through 7 are the Watch-and-Plan zone, and days 6 through 0 are the Act zone.

These quantities are further subdivided into three series groupings of status type based on a priority dispatch scheduling algorithm. The first series group, colored in red, are those units that are late to their scheduled arrival date at the constraint. The second series group, colored in green, comprises units that are on time to their scheduled arrival at the constraint resource. The final series group, colored in blue, consists of units that are early to their scheduled arrival at the constraint resource. The goal is to eliminate the red portions of the graphs so that all products are on schedule. The Constraint Buffer is also broken into three zones. These zones are the Expedite Zone, Watch-and-Plan Zone, and the OK Zone. Zones are indicated by the presence or absence of crosshatch fill patterns on the bar graph. The Expedite Zone includes those units within a 6-day span from the constraint resource and provides a visual cue for operators to take immediate action on any units falling into the red (late) sections of the Expedite Zone.

The Shipping Buffer graph (Figure 7.7) displays the same type of information as described above. The graph relates to the end of the manufacturing flow rather than the constraint point. Product is color coded as late, on time, or early based its expected ship date from the manufacturing facility.

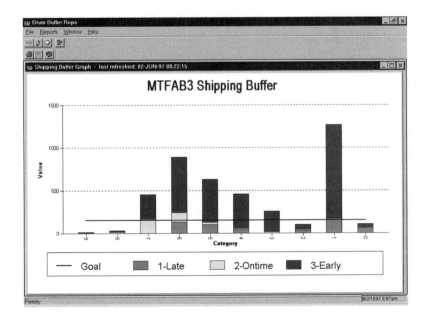

Figure 7.7   DBR scheduling tool—shipping buffer.

Both graphs allow the user of the interface to "drill down" to manufacturing lot detail with a click of the mouse button (Figure 7.8). Information at this level allows the shop floor operators to locate lots in the manufacturing flow quickly and take action to adjust their priority as required to keep them on schedule. The detailed information is configurable at the user's PC, thus allowing the user to hide or display as much information as desired.

This chapter has covered the basics of synchronous flow manufacturing and its production control system, drum-buffer-rope. SFM is a pull system like kanban but it offers two advantages: (1) a stoppage anywhere in the line except at the constraint operation does not affect the factory's throughput, and (2) production control may be simpler because the constraint controls production starts. An inventory buffer protects the constraint against upstream stoppages, and this should be the only work-in-process inventory.

The book has, so far, taken an individual factory to lean manufacturing. The leanest factory cannot, however, achieve its full potential when suppliers and subcontractors deliver or process materials and subassemblies in batches. Long transportation times add cycle time and promote batching, and this aggravates the problem further. The next chapter deals with *supply chain management*, which seeks to overcome these problems.

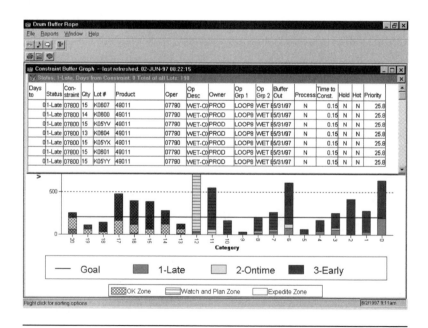

Figure 7.8 Lot details (produced by double clicking a graph bar).

8

Supply Chain Management

Supply chain management extends TOC, JIT, and lean manufacturing techniques to every entity in the supply chain. Walker (2001) defines a supply chain as a global network that delivers products and services from raw materials to end customers through engineered flows of information, physical distribution, and cash. The model treats customers and suppliers as trading partners whose prosperity depends on that of the entire supply chain. It warns against suboptimization, in which each organization tries to maximize its own efficiency and profitability without looking at the overall effects of its actions.

Any business relationship must be a win–win deal for both the customer and the supplier or it cannot last:

> Exact as much as possible at the lowest possible price from your suppliers and carriers of materials; provide the least possible service at the highest price to your customers.
>
> Harsh sounding, but good business? If it ever was, it's not anymore (Schonberger 1986, 155).

Henry Ford made it clear in *Today and Tomorrow* (1926, 40) and his other books that this was never good business. He condemned the practice of paying workers and suppliers as little as they will take while charging customers all that the traffic will bear. The way to reduce labor costs was to make jobs more efficient and the way to reduce material costs was to educate suppliers about lean manufacturing.

Furthermore, the leanest JIT system will not work properly if suppliers deliver large batches. Remember that pig-swallowing is best left to large constrictor snakes. Once the pig is in your digestive system, or production line, you can't process anything else for a while (Figure 8.1).[1] Recall that the lead time is the time between placement of an order and delivery of the product. Long subcontractor lead times (for example, to chrome-plate automobile bumpers) obviate the benefits of internal lead time reductions. Schonberger (1986, 8) warns, "Making more than can be sold is costly and wasteful, and the cost and waste are magnified manyfold as the resulting lumpiness in the demand pattern ripples back through all prior stages of manufacture, including outside suppliers."

This is the importance of supply chain management, without which the leanest factory cannot achieve its full potential. One of the major obstacles to effective supply chain management is suboptimization in supply chains. When each member tries to optimize its own profit without regard for the entire chain's performance, the result is often that everybody makes less money.

Figure 8.1 Large batches undermine the workings of even the best JIT systems.

DYSFUNCTIONAL BEHAVIOR
IN SUPPLY CHAINS

Walker (2001) warns that the largest barrier to successful supply chain management is *lack of trust between trading partners*. It is easy to list some causes. Purchasing departments that try to squeeze concessions from suppliers (or vice versa) damage trust and create dysfunctional responses. The supplier might say, for example, "We'll give you the five percent discount you want if you take 1000 units instead of 500." The purchaser may think, "We will have to keep the 500 extra units in inventory but we'll use them eventually. Meanwhile, I'll look good for getting the five percent discount." Once again, the inventory piles up.

Car purchasing is an activity in which there is no trust between the retailer and the end customer. Car dealers try to pack expensive but worthless options into deals. "Environmental packages" (rustproofing for two or three times its actual value), "fabric treatments" (spraying a $10 can of fabric protector over the seats and charging $100), and ADM (additional dealer markup) are typical examples of game-playing. Conversely, dealers are unhappy when customers also play games. Customers know that new cars lose much of their attractiveness (but not, of course, their mechanical function) when the model year changes. Customers who wait for the model year to end can expect discounts from the dealer and rebates and incentives from the manufacturer. The cars have, meanwhile, run up carrying costs by sitting on the dealer's lot for half a year or so.

Womack and Jones (1996, 82) point out, "Whenever we drive by a car dealer our first thought is always the same: 'Look at all that muda, the vast lot of cars already made which no one wants.' . . . 'Why did the dealer order cars and service parts which aren't needed, and why did the factory build cars and parts in advance of customer pull?'" Making cars to order, as recommended by Levinson (1994, 204, 1998, 296, and 2002) would cut the non-value-adding dealer out of the loop, reduce the customer's cost while increasing the manufacturer's profit, and assure that every car that leaves the production line has a buyer.

One barrier to this is the time needed to fill an order. Many customers are unwilling to wait eight or more weeks for a factory order's completion. Toyota's lean production system can, however, build a car to order in about a week.

Lack of Communication

Womack and Jones (1996, 20) describe waste in the aerospace industry. Pratt & Whitney discovered that machining processes turned 90 percent of specialty metal ingots into chips and only 10 percent into parts. ". . . the initial ingot was poured in a massive size—the melters were certain that this was efficient—without much attention to the shape of the finished parts." The idea of minimizing machining waste by making the ingots or billets in the rough shape of the final part is neither rocket science nor is it new: "Our objective is always to minimize the subsequent machining" (Ford 1926, 69).

Pratt & Whitney, meanwhile, had asked the melters to prepare alloys that differed only slightly, thus increasing costs. The technical requirements for different engine families differed only a little but the designers apparently never considered specifying the same alloy. *Group technology* involves standardizing components to reduce the amount of part numbers. Different product designs should use existing parts and share part numbers if possible, instead of requiring different and possibly new parts for every design. In this example, Pratt & Whitney could have simplified its material requirements by using the same alloy for several engine families. Other advantages of group technology include (Heizer and Render 1991, 273–74):

1. Improved designs, due to design rationalization and elimination of part duplications. (In other words, make the parts interchangeable; Eli Whitney told us this almost two centuries ago.)

2. Lower raw material costs.

3. Simplified production planning and control (because "Machines can process families of parts as a group, minimizing setups, routings, and material handling").

4. Improved routings and machine loadings.

5. Reduced setup times, less work in process (inventory), and shorter production times.

The waste at Pratt & Whitney went unnoticed for decades because the four companies in the supply chain—the melter, forger, machiner, and final assembler—had never taken the time to understand each other's operations. They paid close attention to their own activities but they did not understand how those activities affected the rest of the value stream (Womack and Jones 1996, 20).

The Beer Game: Long Feedback Loops

The Beer Game is a simulation that shows what happens when no one manages the supply chain. References appear in several books and on the Internet.[2] Figure 8.2 shows the supply chain.

Figure 8.2 Beer supply chain.

Each entity's goal is to satisfy its customers and make as much money as possible. Holding down inventory costs is, of course, helpful. Order sheets are the only communication medium in this supply chain. The reader can probably imagine the piles of inventory, unfilled back orders, and dissatisfied end customers.

The problem is characteristic of any physical or business process control system that has long information feedback loops.[3] The overall system begins to respond, not to the actual input conditions (customer demand), but to its own unstable feedback loop. The Beer Game's retailer might, for example, want to keep a one-week stock. If it gets a lot of business during one week—which may represent random variation, not a change in the actual average demand—it sends a larger order to the wholesaler. It may take a while to process this order and the retailer may, upon inspecting its dwindling stocks, send another order. (Remember that order sheets are the only communication medium.)

Now suppose that the retailer's sales fall off—again due to random variation, not because of a change in average demand—perhaps at the time the wholesaler's deliveries come in. Now the retailer is overstocked and it stops placing orders. Remember that these unsteady signals—large orders and/or duplicate orders followed by no orders—are working their way upstream to the distributor and the factory. The resulting mess should be quite easy to imagine.

Suboptimization

Schragenheim (2000) gives an example of suboptimization in which two entities in the supply chain have different profit drivers. Giant Electronics Inc. buys a $500 part from key supplier Tiny Inc. and installs it in a product that sells for $20 million. Giant sells 50 such units a year for revenues of $1 billion.

The $500 part therefore produces annual revenues of $25,000 for Tiny. A modification would triple Giant's consumption of this part over time but Tiny is unwilling to spend time on it. It is focusing instead on another product that will bring in $6 million annually. This leads to animosity between the two companies. Giant's CEO wants the modified part *now*, or at least within the next few months, while Tiny's wants to bring his new product to market. The article assumes that modification of the part for Giant will require one month and delay this new product's introduction for one month. The article asks, "The two executives are supposed to meet in a week. Is there a workable solution to this difficult conflict?"

Schragenheim's key points, which the neighboring article (Holt and Button 2000) underscore, are:

1. "To maximize the revenue of the entire chain, links must make decisions in the interests of the entire chain."

2. "To protect the interests of the links, the chain must make decisions in the interests of the links."

The problem of supply chain management is to resolve the often-inherent conflict between these two objectives.

Chapter 9 will treat linear programming (LP), which can select the product mix that delivers the most profit subject to capacity and other constraints. The supply chain issue brings up the consideration that the LP problem statement might have to span several business entities. There might even be a need to change pricing between supply chain members to increase profits for all. (*Transfer pricing* is often a source of conflict in vertically-integrated organizations that measure each business unit on its profits.)

The following simple example does not require a linear programming solution. Use the marginal profit (profit from selling one additional unit) as the performance metric.

- A subcomponent supplier has 100 arbitrary units of capacity

    - Component A requires 5 units and earns a marginal profit of $100

    - Component B requires 1 unit and earns a marginal profit of $25

- The finished goods manufacturer can:

    - Assemble component A into a product whose marginal profit is $500

    - Assemble component B into a product whose marginal profit is $50

The optimum decision for the subcomponent supplier is to make 100 units of B for $2500 profit. This, however, allows the finished goods manufacturer to earn only $5000. Production of 20 units of A would earn $10,000 for the manufacturer. The supplier is unlikely to be receptive to this idea because it earns only $2000.

Suppose these two companies look at the total profit picture. Making 100 units of B earns this two-entity supply chain $7500. Making 20 units of A earns $12,000. $12,000 in profit is obviously better than $7500. This is achievable if the finished goods manufacturer gives the subcomponent supplier at least an extra $500 for making A instead of B.

Make it $1000 instead and examine the results in Table 8.1. The left figure is the outcome for the supplier and the right figure is for the manufacturer.

Note the similarity to the prisoner's dilemma and the guns-versus-butter dilemma, which also encourage suboptimal decisions. The key premise is that the two parties cannot communicate or cooperate (or, if they can

Table 8.1 Cooperation versus competition.

Supply Chains

		Supplier	
		Cooperate	**Suboptimize**
Manufacturer	**Cooperate**	$3000/$9000	$2500/$5000
	Suboptimize	$2000/$10000	$0/$0 (no deal)

Prisoners' Dilemma

		Prisoner A	
		Don't confess	**Confess, testify against B**
Prisoner B	**Don't confess**	Acquitted/Acquitted (due to lack of evidence)	1 year/10 years in prison (A gets leniency for testifying against B)
	Confess, testify against A	10 years/1 year in prison (B gets leniency for testifying against A)	3 years/3 years in prison (both having pleaded guilty)

Guns versus Butter

		Country A	
		Buy butter	**Buy guns**
Country B	**Buy butter**	+3/+3 (both country's citizens have plenty of butter)	+4/−4 (A conquers B, takes B's butter)
	Buy guns	−4/+4 (B conquers A, takes A's butter)	0/0 (A and B have peace through mutual deterrence but no butter)

communicate, they do not trust each other). This brings us back to Walker's (2001) point that trust is a vital element in successful supply chain management.

The importance of clear communications in supply chains applies to both order placement (for example, the Beer Game) and product requirements (such as ingot sizes and compositions). Suboptimization results in less profit for everyone, including the supply chain elements that think they're doing the best job for themselves. The next consideration in supply chain management is understanding the value stream. This includes identification of value-adding and non-value-adding activities.

UNDERSTAND THE VALUE STREAM

Womack and Jones (1996, chapter 2) describe the steps that result in the delivery of a can of soft drink to the British Tesco grocery chain. The aluminum can, incidentally, accounts for *half the product's cost*. Half the product is waste as delivered because the customer obviously does not drink the can.

The can's cost is not surprising. It begins as bauxite, or aluminum ore. A load of bauxite might travel from Australia to Scandinavia (waste or muda of transportation). The idea of making the aluminum in Australia and thus not transporting the oxygen that comprises half the ore's weight comes to mind.[4] The problem is that economical aluminum manufacture depends on cheap electrical power. Scandinavia has abundant and cheap hydroelectric power, Australia does not. The relative electric power costs therefore outweigh the transportation costs. The aluminum is then fabricated into cans.

Meanwhile, corn is grown and converted into caramel, and beets are made into sugar. The authors show that the drink is actually mixed at the bottler, who also fills the cans and packages the cans in cartons.

Womack and Jones continue, "From the can maker's warehouse, the cans are trucked to the bottler's warehouse . . . " The can maker or the bottler (depending on who pays for delivery) pays a truck to carry a shipment that is mostly air to a warehouse. The warehouse then stores empty cans (air) until they are used. Schonberger (1982, 237) notes that Anheuser-Busch, Inc., receives and uses an almost-continuous stream of truckloads of empty cans. It is easy to make two recommendations:

1. Ship the drink in a full tank truck to the can maker and fill the cans as they are manufactured

2. Use plastic instead, and blow-mold the plastic bottles at the point of use: the bottler

The latter concept also applies to packing materials like foam peanuts. They are mostly air and their transportation is waste. This waste can be avoided by making the packaging foam at the point of use.

We have shown just how much waste can work its way into the value stream and how its elimination can result in more profit for the supply chain and lower prices for the consumer. The next section shows that, for ideal results, the entire supply chain must have just-in-time capability. Individual entities cannot achieve their full potential if their suppliers do not have JIT capability.

MAKE THE SUPPLIER AN EXTENSION OF THE FACTORY

SFM and JIT work even better if suppliers can deliver small lots frequently. "In the delivery area, frequent, small lot, on-time deliveries should be targeted in order to make the manufacturer-supplier linkages tighter" (Suzaki 1987, 197). This goes against the traditional approach of delivering large lots to fill the entire truck or rail car. It has the advantages of:

1. Reducing inventory in transit, or "float"

2. Reducing the need for storage space for deliveries

3. Allowing earlier detection of quality problems, with rapid feedback to suppliers

Suzaki (1987, 203) encourages manufacturers to "think of the suppliers as an extension of the factory." Intranets, extranets, and similar computer networks can link the supplier's production scheduling system to the customer. The Weyerhaeuser intranet (chapter 1) was an example.

The customer's constraint can beat the drum for the supplier as well as for internal production starts (Figure 8.3). It is most effective to tie the customer's constraint to the supplier's constraint because the latter controls the delivery rate. The supplier's constraint steps into the role of the customer's production start operation.

Figure 8.4 treats the interfactory production management system as a kanban system. It assumes that the lead time between the supplier's and customer's constraint is short enough to maintain a very small shipping buffer. The customer's constraint requests three finished units of product X and 1 of Z. This tells the supplier's constraint to begin work on three subassemblies or work-in-process units of X, and one of Z. Two of the three X will replace the two units in the shipping buffer, and the third will be shipped to the customer (with or after the two from the shipping buffer). The constraint will also make a unit of Z to replace the one in the shipping buffer.

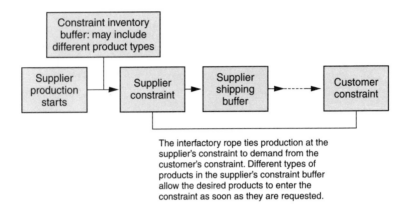

The interfactory rope ties production at the supplier's constraint to demand from the customer's constraint. Different types of products in the supplier's constraint buffer allow the desired products to enter the constraint as soon as they are requested.

Figure 8.3 An interfactory rope can tie the customer's constraint to the supplier's.

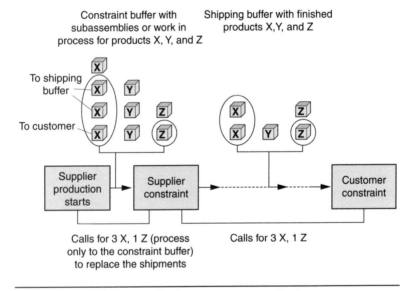

Figure 8.4 Interfactory production management as kanban.

The supplier constraint tells production control to release materials for 3 X and 1 Z. The preconstraint operations will process these to replace the units in the inventory buffer. This is only one example of how this system can be managed.

Throughput and the Supply Chain

A key principle of TOC is that revenues are earned only when a product is sold, not when it is produced. This concept extends throughout the entire supply chain. There is no reason to make a subcomponent, or even buy the materials for it, unless it is destined for a finished product for which there is an order. This is a further strong argument for extending the "pull" logistics system through the entire supply chain.

Low Inventories Reduce Dependence on Market Forecasts

A dependable market forecast is a prerequisite for smooth manufacturing operations. Customers, however, often prefer to place orders only two to three months before they need the items. High inventory levels can lengthen production lead times beyond this forecast horizon. This makes pure guesswork, not reliable forecasts, the basis for production planning.

Smaller inventories shorten production lead times. There is less chance of making the wrong products. The factory's own orders also will be more reliable and less disruptive to its suppliers. "A prime reason that our vendors cannot deliver reliably is because we keep changing our requirements on them, the same way our customers are changing their requirements on us" (Goldratt and Fox 1986, 60). In contrast, coordination of customer and supplier constraints makes the forecasts completely reliable.

We have shown the desirability of linking the production control systems of suppliers and customers. Modern information systems make this much easier. The next section discusses problems that are often inherent in complex supply chains. There is a difference between simplicity and ease. The concepts of supply chain management are simple, yet their implementation is often very challenging.

PRACTICAL PROBLEMS IN COMPLEX SUPPLY CHAINS

The ideal of having the supply chain's constraint beat the drum for every trading partner can, in practice, become a nightmarish proposition.

Simplistic models imply that synchronization extends across the entire breadth of the supply chain. While this may be true for high-volume, low-complexity goods, it cannot be true for low-volume, high-complexity goods. This is because of the complexity found in real supply chains in

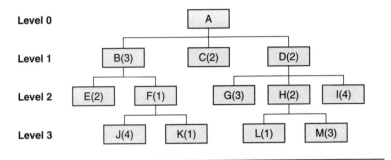

Figure 8.5 An exploded bill of materials (BOM).

terms of the depth of the bills of materials (BOMs), the sheer number of parts suppliers, and the severity of the longest lead times (Walker 2001a).

Figure 8.5 shows the general idea. The exploded BOM shows the requirements for each component and subcomponent. This is a simple example but it is easy to imagine far more levels and far more children (subcomponents in the next level down) per parent. For example, B is the parent of E and F.

It takes three units of B, two of C, and two of D to assemble the product A. B, C, and D may come from different suppliers. This is not too bad so far because the producer's constraint can conceivably link to each supplier's production control system. Recall that Weyerhaeuser's Valley Forge Fine Paper Company's master production schedule responds to customer sales information that comes through an extranet.

It takes 2 units of E and one of F to make B. If B's manufacturer makes these itself, its own JIT or SFM production system should be able to manage affairs. If it has to buy them from suppliers who are still doing batch-and-queue and make-to-forecast, inventory is going to accumulate somewhere—either at B's manufacturer's receiving dock or in a supplier's warehouse. It gets worse if E or F comes from an offshore supplier who was selected due to low labor costs. The time in transit is longer and shippers like to wait for full loads.

Every supplier in this BOM must be reliable. Think "series reliability," in which the failure of even one element causes the entire system to fail. An interruption in the supply of L or M makes it impossible to produce H. This stops production of D, which in turn prevents completion of the final product. This consideration is why manufacturers often like to keep safety stocks (inventory) to decouple them from supply chain disruptions. It's also a reason for having more than one supplier.

SFM says that inventory at the constraint buffer is enough to preclude opportunity losses due to interruptions anywhere in the system. The system is not immune to a protracted stoppage in the supply chain but neither is one that keeps warehouses full of everything; *something* vital will eventually run out. Absolute safety is achievable only by thinking like medieval lords who kept half a year's supply of food in their castles in case a siege cut off their supplies (Figure 8.6).

The modern equivalent of a siege would be a natural disaster or economic collapse that prevented a supplier from delivering. In mid-1993, a fire crippled the Sumitomo plant that makes 60 percent of the world's supply of an encapsulation resin for semiconductor products ("Resin crisis averted," 1993). A 1992 strike at General Motors' Lordstown, Ohio metal stamping plant shut down the Saturn assembly plant, which used a JIT system and had little safety stock, in Tennessee four hours later. This idled 40,000 or so workers for nine days and cost $50 million a day, which probably does not include the opportunity costs of the lost production (Courtney 1992).

Walker (2001a) cites the example of ultra-low-temperature coefficient power resistors that require specialty alloy metals. They come in 4000-pound (1818 kg) ingots that the manufacturer schedules for production every two years.

The bill of materials (BOM) for an automobile can include 10,000 parts. Electronic instruments can have 1800 parts with 16 levels in their product structures. The BOM for such a product can be traced to hundreds of suppliers, not all of whom know the merits of short lead time production. This underscores the following discussion of the importance of supplier

CEO's office

I'm not afraid of my competitors, half of this building is a warehouse

Figure 8.6   The medieval view of inventory and safety stocks.

education. Even one supplier who is out of step with the idea can hobble an otherwise smoothly-functioning supply chain. The effect is similar to that of a single batch-and-queue operation like heat treatment in an otherwise single-unit flow process.

Use Fewer Suppliers

A key advantage of vertical integration, as practiced by Henry Ford, is control of the entire supply chain. Sole-sourcing and supply base reduction offer many of the advantages of vertical integration (regular communication with the suppliers and coordination of activities) without the disadvantages (the need to own the entire company).

Nihon Chukuko, a truck parts manufacturer, gets 90 percent of its business from Isuzu, which in turn has helped develop Chukuko as a supplier. Chukuku delivers according to Isuzu's JIT schedule: 1:00 P.M. ± 30 minutes, and the deliveries are usually incorporated into trucks within two hours.

Per Schonberger (1986, 156), the Xerox Reprographics Division went from 5000 to 300 suppliers and IBM's Typewriter Division went from 640 to 32 suppliers. Suppliers who sell higher volumes to fewer customers don't have to deal with conflicting customer demands. Fewer suppliers mean less of a series-reliability problem in a complex bill of materials such as the one described earlier. It's easier to perform supplier development when one is working with only a few suppliers.

This section has exposed some of the problems that can arise in complex supply chains. The product's bill of materials often shows the proportions that complexity can take. A stoppage anywhere in the supply chain can be catastrophic (once constraint buffers run out) but the only way to achieve total safety is to keep a warehouse full of everything in the BOM, an approach suitable for medieval castles but not modern lean manufacturers. Minimizing the number of suppliers mitigates against the series-reliability problem, and supplier development may make the suppliers more reliable as well as more efficient. The next section treats supplier development in detail.

SUPPLIER DEVELOPMENT

Suppliers and subcontractors cannot be effective trading partners in a lean supply chain unless they master lean techniques themselves. Schonberger (1982, 163) describes supplier development as "missionary work," in which the customer teaches its suppliers lean manufacturing principles like JIT.

Ford (1926, 43–44) provides an example of supplier development from the early 20th century. A supplier was making very low profits on automobile bodies that he sold to the Ford Motor Company. Instead of giving him a higher price, Ford examined his operation and discovered that the bodies should cost half as much. The supplier finally agreed to try to meet this price. This compelled him to improve his business practices. He raised wages to attract the best workers, the workers whom Taylor called "high-priced men." Necessity compelled him to reduce costs wherever possible. The supplier ended up making more profit under the low price than he made under the high price, and his workers received more pay.

Polaroid's zero-base pricing (ZBP) works the same way (Schonberger 1986, 157). Supplier cost increases are not acceptable reasons for the supplier to raise its price. Instead, Polaroid teaches the supplier how to contain costs. The goal of ZBP is to set target prices and show the supplier how to keep costs down.

Porsche's suppliers agreed to provide JIT deliveries—out of their own massive warehouses and inventory stockpiles. Porsche realized that it had to enlighten these suppliers because it bought almost 80 percent of its manufacturing value from them (Womack and Jones 1996, 207–8).

Shalid Khan's Bumper Works learned JIT from a Toyota *sensei* (teacher) but there were still problems. Bumper Works had to send automobile bumpers to Chrome Craft for chrome plating. "Bumpers disappeared into Chrome Craft and didn't reappear for weeks" (Womack and Jones 1996, 71). Khan and the Toyota sensei had to teach Chrome Craft how to perform JIT. The subcontractor's inventory turns on Toyota bumpers increased 2500% (from 20 to 500). Chrome Craft had other customers and the knowledge doubtlessly helped improve productivity with their orders too.

Customer Contact Teams

Customer contact teams (CCTs) are teams of frontline workers who meet with a customer's or supplier's frontline workers. They take advantage of the fact that the frontline worker often knows more about the job, including supplier materials, than anyone else. Sands (in Levinson 1998, 89) describes CCTs at Fairchild Semiconductor's plant in Mountaintop, Pennsylvania as follows.

> . . . Customer Contact Teams (CCT's) promote customer–supplier communications at the shop floor level. Most CCT members are manufacturing workers, and they talk directly to the customer's

frontline workers. CCT's solve problems and deliver customer-specific quality improvements. They rely on three characteristics:

1. They use the frontline manufacturing workers' knowledge, skill, and experience

2. They open short, direct communications between the people who make a product and the people who use it

3. They improve sensitivity toward customer concerns within the organization

Rolm Telecommunications adopted a similar approach. Wayne Mehl, former vice president and general manager, describes the initiative. "We sent some people back to work on their assembly line, to understand their problems. We had some of their people come out and work in ours, putting their product in our products" (Schonberger 1986, 158–59). Figure 8.7 underscores the author's further observation about the ineffectiveness of communication only between the customer's buyer and the supplier's order-entry person.

Customer Development

This is supplier development in reverse. Remember that the nice part about hitting yourself on the head with a hammer is that it feels good when you stop, and Juran (1992, 198) provides the following example:

> Another example of an initiative by the supplier involved a special size of ingot demanded by only one customer. . . . During a visit to the customer, the question was raised: "No one else requires a special size. Why do you need one?" The reply was, "Our storage chest is not big enough to hold the regular size." The supplier then built a bigger storage chest for the customer.

Supplier development, or education of suppliers in lean enterprise techniques, is extremely important because the leanest factory cannot achieve its potential when it works with non-lean suppliers and subcontractors. Suppliers must sometimes educate their customers as well; remember that the supplier's prosperity depends on the customer's success.

Transportation is another element of supply chain management. Japan and Europe have an advantage over North America and Russia because supplier–customer distances are often shorter. Innovative freight management services (FMSs) and third party logistics (3PL) systems can overcome many geographical problems.

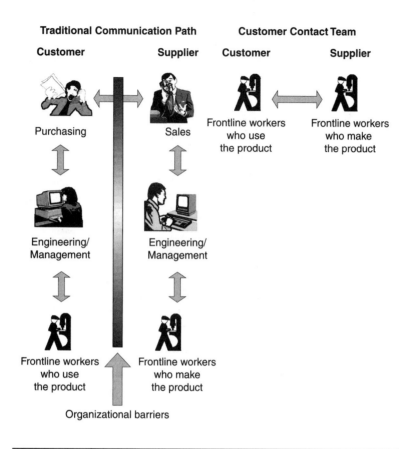

Figure 8.7   Traditional buyer–seller communications versus CCTs.

TRANSPORTATION IN THE SUPPLY CHAIN

Transportation complicates supply chain management. Transportation adds lead time, especially for material that travels on ships. Material in transit, or float, is inventory.

In an ideal JIT world the supplier would be next door and its production control system would be tightly integrated with the customer's. Europe and Japan have high population densities, hence suppliers may be only a couple of hours away by truck or train. Distances are often much greater in North America and this creates problems that must be overcome. Russian companies that try to adopt JIT also will have to address the distance problem.

World Sourcing: Don't (Unless There's a Compelling Reason)

Schonberger (1986, 162) decries the "world-sourcing" fad, in which companies shop the world for the lowest price. The lowest price is subject to foreign exchange rates that change frequently. This section discusses other problems with offshore production.

We have already shown why the loss of manufacturing jobs is dangerous to national prosperity and military security. Walker (2001a) provides an excellent supply chain management argument against continuing to send manufacturing jobs offshore in a quest for lower labor costs: it introduces batching and queuing into what might otherwise be a JIT system. Supplier shipments might have to wait for consolidation into full container loads (big batches) for a container ship. The container ship is an even larger batch-and-queue operation that will want to sail with a full hold. "A supply chain that is completely synchronized downstream within a country or a geographic region will lose its synchronization at some point upstream where the international logistics-driven lot sizing takes effect."

Henry Ford's solution was to own both the ships and the suppliers, so shipments were waiting on the loading docks when the ship arrived to get them. The ship was rarely in port for 24 hours. Even this system, however, couldn't help accumulating inventory because a ship is a batch operation. A system that involves different suppliers who may be members of different supply chains is even worse. They all must get their products to the dock in time to meet the ship. This means that *someone's* shipment will have to wait, either on the dock or in a warehouse. If the customer needs it on a certain date (and there may be several customers with different delivery requirements) the ship must arrive before the earliest date (Table 8.2).

For example, it is about 7295 miles (11740 km) from Los Angeles to Manila in the Philippines, 6490 miles (10440 km) to Shanghai, China, and 5480 miles (8820 km) to Tokyo, Japan.[5] Under the generous assumption that a cargo ship can make a steady 25 knots (45.7 km/hour), it takes at least eight days to travel the shortest distance shown here. In this example, the customer for C is the only customer who can have it "just in time"—and even then the materials are in transit for eight days.

The lead time for any order is also much longer. For example, the customer for A needs it on day 12. It must be at the dock on day 3, which means the customer must order it at day 3 minus the amount of time the supplier needs to produce it and get it to the harbor. Transportation adds nine days of lead time (time between order placement and order delivery). Single-day lead times are realistic for production of certain items, as shown by the Omark example. The need to put the items on a cargo ship can obviate the benefits of even the most responsive supplier's JIT production system.

Table 8.2 Logistics problems with offshore production (a ship is a cleverly-disguised warehouse).

Available Shipping

Departs Asia	Arrives U.S.
Day 3	Day 11
Day 9	Day 17
Day 15	Day 23

Material	Required on:	Order in Advance	Waiting Time in U.S.
A	Day 12	9+ days	1 day
B	Day 16	13+ days	5 days
C	Day 17	8+ days	0 days
D	Day 21	12+ days	4 days

The situation is even worse than described above because lead times are often twice as long:

> For example, in the case of ocean cargo from Asia to the United States, transportation lead-time door-to-door can be in excess of 25 days. In such cases quality must be absolutely reliable. . . . once merchandise enters the logistics chain, reversing the process is very difficult and expensive to handle. JIT operations cannot afford to have compromised shipments in the pipeline, particularly when the problems don't surface until they hit the port of destination (Gardner 2001).

This is the same issue that JIT addresses in a factory: problems can hide in inventory that waits a long time for the next operation. Goldratt and Fox (1986, 62) point out the benefits of dealing with domestic suppliers:

> In some cases, it has been possible to actually command premium prices when quoted lead times are substantially less than other competitors. This is a huge competitive advantage that many Western competitors could have over foreign competitors because of the time required for ocean freight shipments.

Womack and Jones (1996) note that Japan has a geographical disadvantage because it must ship many of its products overseas. When Toyota had no factories in the United States it had to send cars by ship. This added to the delivery cost and also created several days of "float" (materials or subassemblies in transit).

Overseas sources are less troublesome when parts have high value-to-weight ratios. Then can travel economically by air for next-day deliveries.

Semiconductor die or chips are often sent to the Far East (Philippines, Malaysia, Singapore) for assembly into electronic packages and are then sent back. Menlo Tool Company makes small carbide tools that, in 1993, sold for about $100 per pound ($220 per kg). The company president said, "We can put 60 pounds [27.3 kg] of product—worth $6000—into a 14 by 12 by 8 inch [35.6 by 30.5 by 20.3 cm] box, give it to any air-freight carrier, and ship it quickly and economically overseas overnight. . . . If we have it in inventory here and a competitor over there does not, we will beat them every time" (Kastelic 1993, 104). In a JIT system, of course, there would be only a small finished goods buffer of tools, and products would be made to order.

Domestic JIT: Truck Sharing

Distance is a major objection to daily or semi-daily just-in-time deliveries in the United States. Europe and Japan have an advantage here because of their higher population densities but the problem is not insurmountable.

Schonberger (1986, 167) shows that one so-called solution has been to set up JIT warehouses. Suppliers deliver full truckloads, which might comprise a week's or a month's supply, to the warehouse. The JIT warehouse then makes daily or sub-daily deliveries according to the customer's needs. "Just in time" and "warehouse" are, of course, contradictory terms (Figure 8.8).

It is not economical to send a daily truck with, for example, a ten percent load from the supplier to the customer (Figure 8.9). (There are, however, apparently companies that will fill a truck with empty aluminum cans

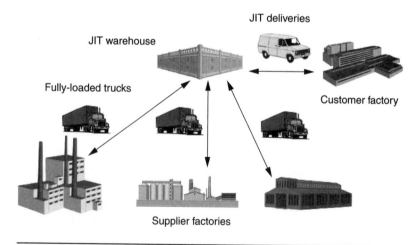

Figure 8.8   The JIT warehouse: an oxymoron.

or glass bottles and call it a full load.) Cargo ships and trucks invite batching for the same reason: they minimize the cost per unit of weight.

Truck sharing, not JIT warehousing, is the obvious answer. A daily or even a sub-daily truck circulates among suppliers and customers like a bus (Figure 8.10).

There are well-defined operations research algorithms for finding the optimum route. In the transportation problem, there are N sources, each of which has a certain capacity, and M destinations, each of which has a certain requirement. C_{ij} is the cost to make one shipment from source i to destination j. The transportation simplex algorithm finds the lowest-cost solution (Hillier and Lieberman 1980, chapter 4). The related *transshipment problem* extends the transportation problem to cover intermediate transfer points. These can be junctions, sources, or destinations. It should be possible to formulate a system of N suppliers (sources) and M customers (destinations) as a transshipment problem, for which there are well-established optimization routines.

Schonberger (1986, 168) describes the location of automotive parts suppliers in Kearney and Lincoln, Nebraska, and Des Moines, Iowa. These are 923, 793, and 604 miles (1485, 1276, 972 km) respectively from the customer in Detroit, Michigan. There is no reason why a truck cannot pick up pistons from Eaton Corporation in Kearney at 05:00, V-belts and hoses from Goodyear in Lincoln at 08:30, other parts in Des Moines at 13:00, and have them in Detroit in time for the next day's morning shift. (24-hour or military times allow easy calculation of the travel times.)

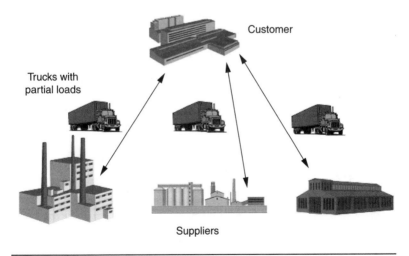

Figure 8.9 This transportation model doesn't work well either.

What about the problem of the loading and unloading sequence? This is a problem only under the self-limiting paradigm that a truck must load and unload from the rear. Fruehauf has developed a gull-winged trailer that can open on both sides. This makes it possible to unload a full shipment of diesel engines in fifteen minutes, an activity that might require a couple of hours with a conventional trailer (Schonberger 1986, 166).

Figure 8.11 suggests yet another idea for having genuine JIT while shipping full loads. The factories in an industrial park (or city) share a small truck that collects everyone's small (JIT-quantity) shipments to a

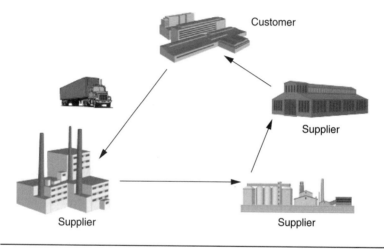

Figure 8.10 Truck sharing.

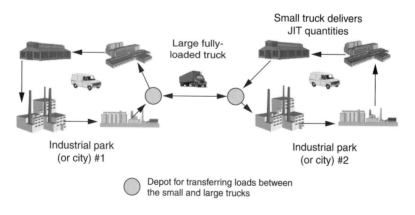

Figure 8.11 Trucking network.

depot for consolidation into a full load for a large truck. The small truck also picks up everyone's small deliveries from the depot.

The figure shows only two clusters of factories but there is no reason why the large truck cannot drop off or pick up shipments from other clusters along its route. A practical obstacle may be incompatibility of shipments but this consideration usually arises with chemicals. There shouldn't be a problem with having plastic, ceramic, and metal parts plus semiconductor wafers (from several different manufacturers, for delivery to several different customers) in the same large truck.

Freight Management Services (FMSs) and Third Party Logistics (3PL)

The preceding section's idea is not particularly obscure. The same or similar strategies are in use by *freight management services* (FMSs). "A freight management service is a systems-based method of managing transportation requirements for greater performance and cost efficiency" (Carter 2001). FMSs reduce shipping costs and also increase profitability for the carriers, for example by end-to-end load matching. This means that when a truck delivers a shipment it can pick up a new load at the same dock instead of traveling empty.

An FMS is expensive—on the order of $2.5 million to purchase, customize, and implement—so many shippers choose *third party logistics* (3PL) services instead. Carter (2001) continues, "The most notable of these benefits may be freight optimization, the process of analyzing and consolidating shipments and leveraging all opportunities to move them via higher-volume, lower-cost transportation modes." These include small package carriers like UPS, truckload carriers, and even rail carriers.

The FMS or 3PL service does not necessarily own the carriers; it may instead orchestrate their activities for maximum value. C. H. Robinson Worldwide Inc. states:

> CHRW is a non-asset-based transportation provider. Because we don't own transportation assets, we can be flexible in finding the best mode and service for our customers. To do this, we use our established relationships with motor carriers, air freight carriers, railroads (primarily intermodal service providers), and ocean carriers. In fact, we have the largest carrier network in North America.[6]

Their Web page underscores the company's role in a supply chain. The company also offers cross-docking, or moving freight across a dock for consolidation for shipment, or deconsolidation (also known as "pool delivery") for delivery to final destinations.

Carter (2001) cites i2 and Manugistics as sources of optimization software. Per a news release at www.i2.com, "FreightMatrix will offer shippers, carriers, and logistics providers with the needed services to buy and sell transportation more efficiently, plan their cargo requirements, and execute the delivery of shipments." Manugistics' NetWorks software performs functions such as "Simultaneously optimizing inbound, outbound, and inter-facility movements," as well as shipment tracking.[7]

An Internet search for "freight management service" and "third party logistics" yields many FMSs. The British Freightwatch company says:

> . . . we will endeavour to consolidate different customers' consignments into the same or similar routings wherever possible, subject of course to any confidential or practical issues that may arise. This is not just restricted to freight purchases, but will also embrace other elements, such as collections, warehousing, and packing. Wherever the opportunity arises we will pass on the benefits.[8]

The acronym LTL (less than load) appears on some of these Web sites, and this shows that FMSs and 3PLs are well aware of this requirement.

This section has shown that transportation invites batching, for example, to achieve full truckloads. Overseas sourcing to take advantage of cheap labor is even worse because the issue now involves full shiploads (or at least full shipping containers). Overseas transportation by ship can pack weeks into cycle times. Innovative transportation systems can, however, overcome these problems. Henry Ford was able to do this in the early 20th century. FMSs and 3PLs that use modern information management systems should make it even easier for us to follow his advice to put brains into the business. We can have cost-effective JIT-quantity shipments despite longer transportation distances in the United States.

ENDNOTES

1. Standard and Davis (1999, 111–12) mention the "pig in a python" effect with respect to large inventory bubbles that move through a factory. "If smaller orders are released more often, the factory resources are loaded much more easily. . . . This is analogous to the python swallowing dozens of little piglets instead of one large pig. . . . Surprisingly, many factories prefer to 'stretch the python' so it can swallow an even larger hog!" The authors go on to cite an explorer who saw an anaconda swallow an adult hog in a Peruvian rain forest. (One of the authors, Dale Davis, studied anthropology and specialized in Mayan culture.) Levinson (1998, 162–64) uses the term "pig-swallowing" and includes a couple of photos of a boa constrictor with a bulge from a recent meal.

2. http://www.syslab.ceu.hu/~workshop/BeerGame/sld001.htm, as of 4/18/01.
3. Chemical engineers will be familiar with this concept from feedback process control. Long lags in feedback make control more difficult and can lead to unstable (or even runaway) situations.
4. Aluminum oxide, Al_2O_3, contains 53 percent aluminum and 47 percent oxygen by weight. This mineral complexes with water, however, so it can contain less than fifty percent aluminum in its raw form. Sienko and Plane (1974, 567–68) provide some technical details on aluminum manufacture.
5. http://www.indo.com/distance/ .
6. http://www.chrobinson.com/cust_serv.asp as of 5/8/01.
7. http://www.manugistics.com/solutions/networks_transport.asp as of 5/8/01.
8. http://www.freightwatch.co.uk/freightsave/ as of 5/8/01.

9

Maximizing Profit in a Constrained Process

This chapter shows how to select the product mix that maximizes profit when the factory is working at its full capacity. Linear programming solves optimization problems that are subject to constraints:

1. Select the product mixture that returns the most money subject to constraints on workstation time. Model "what-if" scenarios such as an increase in capacity at a workstation or elimination of a process step.

2. Do (1), subject also to minimum requirements for certain products, such as preexisting contracts or delivery commitments. Profit optimization must be subordinated to customer satisfaction and contractual duties.

3. Do (1) or (2), subject also to market constraints on production. The market, for example, might absorb only a limited quantity of one or more products.

Linear programming can also identify constraints in multiproduct factories while showing where to add resources to achieve the greatest profits. It can model "what–if" scenarios like changes in resource requirements and sales of products at a discount. Profit optimization subject only to capacity constraints is the most straightforward application.

PROFIT OPTIMIZATION SUBJECT TO CAPACITY CONSTRAINTS

Consider a factory that can make products P, Q, and R. The products have the following characteristics. Each machine is available 20 hours a day (1200 minutes). Table 9.1 shows the time that each product needs at each tool.

Table 9.1   Product characteristics.

			P	Q	R
		Product Name	P	Q	R
		Marginal Profit	8	4	5
Resource	**Type**	**Available**	**Required**	**Required**	**Required**
Blanking	<	1200	4	6	8
Drilling	<	1200	8	4	6
Grinding	<	1200	4	8	4
Heat treatment	<	1200	10	5	8

This table is, in fact, a Microsoft Excel spreadsheet that is readable by a Visual Basic LP simplex program. The "Type" column identifies the type of constraint. A "<" means that the total consumption of the resource must be less than or equal to the amount available, while "=" forces consumption to be exactly equal. (It's also possible to include a greater-than or equal requirement but this does not appear necessary for these applications.)

This example is fairly straightforward. Product P has the highest marginal profit. Heat treatment is obviously the constraint, and it limits production to 120 units for $960.

The solution procedure is as follows. Phrase the problem,

$$
\begin{aligned}
Z - 8x_1 - 4x_2 - 5x_3 & = 0 \\
4x_1 + 6x_2 + 8x_3 + x_4 & = 1200 \text{ Blanking} \\
8x_1 + 4x_2 + 6x_3 \quad + x_5 & = 1200 \text{ Drilling} \\
4x_1 + 8x_2 + 4x_3 \qquad + x_6 & = 1200 \text{ Grinding} \\
10x_1 + 5x_2 + 8x_3 \qquad\quad + x_7 & = 1200 \text{ Heat treatment}
\end{aligned}
$$

where Z is the profit, x_1, x_2, x_3 are production levels for products P, Q, and R, and x_4, x_5, x_6, and x_7 are *slack variables* for the four operations. *The slack variable quantifies the excess capacity at the workstation. A workstation with no slack is a constraint.* (A similar concept appears in project planning where the *sequence of activities with no slack time* is the *critical path*.)

Linear Programming: Simplex Method

Now set up the simplex tableau for this equation system. Copy the coefficients from the equation set into the table. The right-hand side (RHS), incidentally, is the current value of the basic (nonzero) variable. Nonbasic variables, which do not appear in the table, are zero.

The starting basic variables are the slack variables. They tell us that we have 1200 minutes available on each machine. The RHS for Z is the profit. Since the production variables start out at zero there is no profit in the initial tableau.

The LP simplex procedure now allocates the available resources to yield the greatest possible profit. The procedure is (Hillier and Lieberman 1980, chapter 2):

1. Identify the *entering basic variable* by selecting the variable (which will always be a nonbasic variable—a variable that equals zero) with the largest negative coefficient in the Z row. This is the nonbasic variable that increases Z the most; think of the negative coefficient as $-dZ/dx$. This variable's column is the *pivot column*.

2. Determine the *leaving basic variable*—the one the entering basic variable will replace—by dividing the right-hand side by the coefficients in the pivot column. (Do this only for positive coefficients.) The row that yields the smallest result is the leaving basic variable. In the case of a tie, either may be used. In Table 9.2, select x_7 (slack #4) because $1200 \div 10 = 120$, which is less than $1200 \div 4$ or $1200 \div 8$.

3. Divide the pivot row by the pivot coefficient to make the coefficient equal to 1 (Table 9.2a). Then use matrix row operations to drive all the other elements of the pivot column to zero (Table 9.2b). Return to (1) and continue.

Table 9.2 Simplex Tableau, with pivot column and pivot row highlighted.

Tableau—Initial

Basic	EQ	Z	x_1	x_2	x_3	Slack 1	Slack 2	Slack 3	Slack 4	RHS
Z		1	-8	-4	-5	0	0	0	0	0
Slack 1	1	0	4	6	8	1	0	0	0	1200
Slack 2	2	0	8	4	6	0	1	0	0	1200
Slack 3	3	0	4	8	4	0	0	1	0	1200
Slack 4	4	0	10	5	8	0	0	0	1	1200

Table 9.2a Divide the pivot row by the pivot element.

Tableau—Initial

Basic	EQ	Z	x_1	x_2	x_3	Slack 1	Slack 2	Slack 3	Slack 4	RHS
Z		1	-8	-4	-5	0	0	0	0	0
Slack 1	1	0	4	6	8	1	0	0	0	1200
Slack 2	2	0	8	4	6	0	1	0	0	1200
Slack 3	3	0	4	8	4	0	0	1	0	1200
Slack 4	4	0	1	0.5	0.8	0	0	0	1	120

Table 9.2b   Use matrix row operations to make the rest of the pivot column zero.

Tableau—1

Basic	EQ	Z	x_1	x_2	x_3	Slack 1	Slack 2	Slack 3	Slack 4	RHS
Z	0	1	0	0	1.4	0	0	0	0.8	960
Slack 1	1	0	0	4	4.8	1	0	0	-0.4	720
Slack 2	2	0	0	0	-0.4	0	1	0	-0.8	240
Slack 3	3	0	0	6	0.8	0	0	1	-0.4	720
x_1	4	0	1	0.5	0.8	0	0	0	0.1	120

4. Stop when all coefficients in the Z row (equation zero) are nonnegative. Since the coefficients imply $-dZ/dx$, nonnegativity means that further iteration cannot produce further improvement.

Stop because all the coefficients in equation (0) are nonnegative. The results can be read from the right side.

- Profit equals $960 (Z)

- Blanking has 720 minutes of slack, or excess capacity (slack 1 = x_4)

- Drilling has 240 minutes of slack, or excess capacity (slack 2 = x_5)

- Grinding has 720 minutes of slack, or excess capacity (slack 3 = x_6)

- Heat treatment, the constraint, has no excess capacity (slack 4 = x_7). Note that x_7 is a nonbasic variable. It does not appear in the left-hand column of basic variables and it equals zero.

Shadow Prices: Where Best to Increase Capacity?

A resource's *shadow price* is the coefficient of its slack variable in the Z row in the solution. It is the resource's *marginal value*, or the benefit of increasing the resource. In TOC, *the shadow price is the marginal or differential benefit of elevating a constraint* or, in general, providing more of a resource. When the slack variable is positive its shadow price is zero because there is already a surplus of the resource.

There is no benefit to increasing the capacities of blanking, drilling, or grinding. The solution has already shown that they have excess capacity. Ten more minutes of heat treatment capacity would, however, add $8 of profit. *The shadow price tells the factory where to focus total productive maintenance, single-minute exchange of die, and similar efforts to increase capacity.*

Table 9.2c Shadow prices.

Tableau—1

Basic	EQ	Z	x_1	x_2	x_3	Slack 1	Slack 2	Slack 3	Slack 4	RHS
Z	0	1	0	0	1.4	0	0	0	0.8	960
Slack 1	1	0	0	4	4.8	1	0	0	-0.4	720
Slack 2	2	0	0	0	-0.4	0	1	0	-0.8	240
Slack 3	3	0	0	6	0.8	0	0	1	-0.4	720
x_1	4	0	1	0.5	0.8	0	0	0	0.1	120

Table 9.3 New product characteristics: Q and R are now more profitable.

		Product Name	P	Q	R
		Product Value	8	6	6
Resource	Type	Available	Required	Required	Required
Blanking	<	1200	4	6	8
Drilling	<	1200	8	4	6
Grinding	<	1200	4	8	4
Heat treatment	<	1200	10	5	8

Remember that elevation of a constraint will eventually move the constraint somewhere else. The shadow price is the differential or marginal improvement over the current condition. The fact that adding 10 minutes of heat treatment capacity gains $8.00 does not necessarily mean that another 1000 minutes will mean an $800.00 improvement. In fact, adding 301 minutes of heat treatment moves the constraint to drilling.

Example that Requires a Product Mixture for Optimum Profit

The previous example was not particularly interesting except as an exercise to show the parts of the simplex tableau and the information they contain. Table 9.3 shows a more challenging situation in which the highest profit comes from a product mixture.

$$
\begin{aligned}
Z - 8x_1 - 6x_2 - 6x_3 &= 0 \\
4x_1 + 6x_2 + 8x_3 + x_4 &= 1200 \text{ Blanking} \\
8x_1 + 4x_2 + 6x_3 + x_5 &= 1200 \text{ Drilling} \\
4x_1 + 8x_2 + 4x_3 + x_6 &= 1200 \text{ Grinding} \\
10x_1 + 5x_2 + 8x_3 + x_7 &= 1200 \text{ Heat treatment}
\end{aligned}
$$

Table 9.3a   LP Simplex solution, with pivot rows and columns highlighted.

Tableau—Initial

Basic	EQ	Z	x_1	x_2	x_3	Slack 1	Slack 2	Slack 3	Slack 4	RHS
Z		1	-8	-6	-6	0	0	0	0	0
Slack 1	1	0	4	6	8	1	0	0	0	1200
Slack 2	2	0	8	4	6	0	1	0	0	1200
Slack 3	3	0	4	8	4	0	0	1	0	1200
Slack 4	4	0	10	5	8	0	0	0	1	1200

Tableau—1

Basic	EQ	Z	x_1	x_2	x_3	Slack 1	Slack 2	Slack 3	Slack 4	RHS
Z		1	0	-2	0.4	0	0	0	0.8	960
Slack 1	1	0	0	4	4.8	1	0	0	-0.4	720
Slack 2	2	0	0	0	-0.4	0	1	0	-0.8	240
Slack 3	3	0	0	6	0.8	0	0	1	-0.4	720
x_1	4	0	1	0.5	0.8	0	0	0	0.1	120

Tableau—2

Basic	EQ	Z	x_1	x_2	x_3	Slack 1	Slack 2	Slack 3	Slack 4	RHS
Z		1	0	0	0.67	0	0	0.33	0.67	1200
Slack 1	1	0	0	0	4.27	1	0	-0.67	-0.13	240
Slack 2	2	0	0	0	-0.4	0	1	0	-0.8	240
x_2	3	0	0	1	0.13	0	0	0.17	-0.07	120
x_1	4	0	1	0	0.73	0	0	-0.08	0.13	60

The best results come from making 60 P (x_1) and 120 Q (x_2) for $1200. Grinding and heat treatment are both constraints. There are 240 minutes of excess capacity for blanking and drilling (slack variables 1 and 2).

Consider the effect of an additional 10 minutes of heat treatment capacity (slack variable 4). It is now possible to make (theoretically, since a factory can't make fractional units) 61.33 P and 119.33 Q, for $1206.67, or a gain of $6.67. In other words,

$$\frac{dZ}{d(heat\ treatment)} = \frac{\$6.67}{10\ min} = \$0.67.$$

The shadow price is the *marginal profit change per capacity increment*.

Now look at the shadow price for slack variable 3 (grinding). Make 59.17 P and 121.67 Q for $1203.33, a gain of $3.33:

$$\frac{dZ}{d(grinding)} = \frac{\$3.33}{10 \text{ min}} = \$0.33.$$

Effect of a Process Improvement

LP simplex makes it very easy to model "what–if" situations, including the effects of process improvements. Table 9.1 shows that product P offers the most marginal profit but it also requires a lot of heat-treatment capacity. An alloy that does not need heat treatment (following the Omark example) is available but it costs an additional dollar per unit. The factory can make $960 a day in profit under the conditions of Table 9.1. How much can it make if it adopts the new alloy? Simply change P's marginal profit from $8.00 to $7.00 to reflect the more expensive alloy, and also change P's heat-treatment requirement to zero.

Make 100 P and 100 Q for $1100 in profit. This is better than the original situation so the new alloy should be adopted. There are now 200 minutes of excess capacity at blanking (slack #1) and 700 at heat treatment (slack #4).

More drilling capacity (the shadow price for slack #2 is 83 cents per minute) would now be helpful. Another 10 minutes of drilling allows production of 101.67 P and 99.17 Q for $1108.33, an $8.33 gain, or 83 cents per minute.

The next section treats the situation in which the factory must produce a less profitable product to meet contractual or other commitments. LP simplex can address this requirement by adding an equality constraint.

Table 9.4 New alloy for product P.

		Product Name	P	Q	R
		Marginal profit	7	4	5
Resource	Type	Available	Required	Required	Required
Blanking	<	1200	4	6	8
Drilling	<	1200	8	4	6
Grinding	<	1200	4	8	4
Heat treatment	<	1200	0	5	8

Table 9.4a   LP simplex solution for the new alloy.

Tableau—Initial

Basic	EQ	Z	x_1	x_2	x_3	Slack 1	Slack 2	Slack 3	Slack 4	RHS
Z		1	-7	-4	-5	0	0	0	0	0
Slack 1	1	0	4	6	8	1	0	0	0	1200
Slack 2	2	0	8	4	6	0	1	0	0	1200
Slack 3	3	0	4	8	4	0	0	1	0	1200
Slack 4	4	0	0	5	8	0	0	0	1	1200

Tableau—1

Basic	EQ	Z	x_1	x_2	x_3	Slack 1	Slack 2	Slack 3	Slack 4	RHS
Z		1	0	-0.5	0.25	0	0.88	0	0	1050
Slack 1	1	0	0	4	5	1	-0.5	0	0	600
x_1	2	0	1	0.5	0.75	0	0.13	0	0	150
Slack 3	3	0	0	6	1	0	-0.5	1	0	600
Slack 4	4	0	0	5	8	0	0	0	1	1200

Tableau—2

Basic	EQ	Z	x_1	x_2	x_3	Slack 1	Slack 2	Slack 3	Slack 4	RHS
Z		1	0	0	0.33	0	0.83	0.08	0	1100
Slack 1	1	0	0	0	4.33	1	-0.17	-0.67	0	200
x_1	2	0	1	0	0.67	0	0.17	-0.08	0	100
x_2	3	0	0	1	0.17	0	-0.08	0.17	0	100
Slack 4	4	0	0	0	7.17	0	0.42	-0.83	1	700

PROFIT MAXIMIZATION SUBJECT TO CONTRACTUAL COMMITMENTS

The previous solutions did not include product R, which means it is not profitable relative to other opportunities. Now suppose that the factory must deliver 60 Rs per day to a customer because it has a contract or because it doesn't want to disappoint that customer. The constraint, "manufacture exactly 60 units of R" can be built into the problem statement (Table 9.5).

This forces production of R to equal 60 units. Note that ">" could be used to allow production of even more R but this product has already been shown to be unprofitable. The "greater-than or equal" requirement makes the problem solution a little more complicated.

Table 9.5   Product characteristics including a contractual commitment.

		Product Name	P	Q	R
		Marginal profit	8	4	5
Resource	**Type**	**Available**	**Required**	**Required**	**Required**
Blanking	<	1200	4	6	8
Drilling	<	1200	8	4	6
Grinding	<	1200	4	8	4
Heat treatment	<	1200	10	5	8
Commitment	=	60	0	0	1

Linear programming for equality and greater-than constraints requires the use of artificial variables and the "Big M" penalty variable. M is considered infinite so it takes precedence over all other numbers. This forces the solution to meet an equality constraint. In this example,

$$\begin{aligned}
Z - 8x_1 - 4x_2 - 5x_3 \qquad\qquad + M\bar{x}_8 &= 0 \\
4x_1 + 6x_2 + 8x_3 + x_4 \qquad\qquad &= 1200 \text{ Blanking} \\
8x_1 + 4x_2 + 6x_3 \qquad + x_5 \qquad\qquad &= 1200 \text{ Drilling} \\
4x_1 + 8x_2 + 4x_3 \qquad\qquad + x_6 \qquad &= 1200 \text{ Grinding} \\
10x_1 + 5x_2 + 8x_3 \qquad\qquad\qquad + x_7 \quad &= 1200 \text{ Heat treatment} \\
0x_1 + 0x_2 + x_3 \qquad\qquad\qquad + \bar{x}_8 &= 60 \text{ Contractual requirement}
\end{aligned}$$

$\bar{x}_8$ is the artificial variable

An initialization step is required to make the Z-row coefficients of all basic variables equal to zero in the initial tableau. Subtract M times equation (5) from equation (0).

Table 9.5a   Simplex tableau, with equality constraint.

Tableau—Initial

Basic	EQ	Z	x_1	x_2	x_3	x_4	x_5	x_6	x_7	Artif.	RHS
Z	0	1	-8	-4	-5	0	0	0	0	M	0
x_4	1	0	4	6	8	1	0	0	0	0	1200
x_5	2	0	8	4	6	0	1	0	0	0	1200
x_6	3	0	4	8	4	0	0	1	0	0	1200
x_7	4	0	10	5	8	0	0	0	1	0	1200
Artif.	5	0	0	0	1	0	0	0	0	1	60

Note the penalty value, -60M, in the profit row. This will force the solution to include 60 units of product R. Since M takes precedence over all other numbers, column x_3 is the obvious pivot column in Table 9.5b.

Make 72 P (x_1) and 60 R (x_3) for $876. There is excess capacity everywhere but heat treatment. The 80-cent shadow price there is very straightforward. There is excess capacity in the other operations so another 10 minutes of heat treatment allows production of another P for $8.00.

Table 9.5b Simplex tableau with Big M.

Tableau—Initial

Basic	EQ	Z	x_1	x_2	x_3	x_4	x_5	x_6	x_7	Artif.	RHS
Z		1	-8	-4	-M -5	0	0	0	0	0	-60M
x_4	1	0	4	6	8	1	0	0	0	0	1200
x_5	2	0	8	4	6	0	1	0	0	0	1200
x_6	3	0	4	8	4	0	0	1	0	0	1200
x_7	4	0	10	5	8	0	0	0	1	0	1200
Artif.	5	0	0	0	1	0	0	0	0	1	60

Table 9.5c Simplex solution.

Tableau—1

Basic	EQ	Z	x_1	x_2	x_3	x_4	x_5	x_6	x_7	Artif.	RHS
Z		1	-8	-4	0	0	0	0	0	M +5	300
Slack 1	1	0	4	6	0	1	0	0	0	-8	720
Slack 2	2	0	8	4	0	0	1	0	0	-6	840
Slack 3	3	0	4	8	0	0	0	1	0	-4	960
Slack 4	4	0	10	5	0	0	0	0	1	-8	720
x_3	5	0	0	0	1	0	0	0	0	1	60

Tableau—2

Basic	EQ	Z	x_1	x_2	x_3	x_4	x_5	x_6	x_7	Artif.	RHS
Z		1	0	0	0	0	0	0	0.8	M-1.4	876
Slack 1	1	0	0	4	0	1	0	0	-0.4	-4.8	432
Slack 2	2	0	0	0	0	0	1	0	-0.8	0.4	264
Slack 3	3	0	0	6	0	0	0	1	-0.4	-0.8	672
x_1	4	0	1	0.5	0	0	0	0	0.1	-0.8	72
x_3	5	0	0	0	1	0	0	0	0	1	60

These sections have examined the situation in which the constraint is in the factory. The next one deals with market (demand) constraints.

PROFIT MAXIMIZATION SUBJECT TO MARKET CONSTRAINTS

This section shows how to account for market constraints, or limits on the quantity of each product that can be sold. Table 9.6 shows a situation in which the market for P, the company's most profitable product, is limited. The problem formulation treats "demand for P" as a limited resource.

$$
\begin{aligned}
Z - 8x_1 - 4x_2 - 5x_3 &= 0 \\
4x_1 + 6x_2 + 8x_3 + x_4 &= 1200 \text{ Blanking} \\
8x_1 + 4x_2 + 6x_3 \quad + x_5 &= 1200 \text{ Drilling} \\
4x_1 + 8x_2 + 4x_3 \qquad + x_6 &= 1200 \text{ Grinding} \\
10x_1 + 5x_2 + 8x_3 \qquad\quad + x_7 &= 1200 \text{ Heat treatment} \\
x_1 + 0x_2 + 0x_3 \qquad\qquad + x_8 &= 40 \text{ Demand for P}
\end{aligned}
$$

Make 40 P (all that the market will buy), 116.36 Q, and 27.27 R for \$921.82. There are 123.64 minutes of excess capacity at blanking and 250.91 at drilling (x_4, x_5).

The \$1.91 shadow price for P shows a \$1.91 gain for every additional order for this product. If the demand rises to 41, make 41 P, 116.55 Q, and 25.91 R for \$923.73, a gain of \$1.91. It is not the point of this exercise to plan specific production numbers (116.55 Q and 25.91 R) for this situation, especially since the factory cannot make fractional products, but rather to show the sales and marketing departments where to look for orders.

Table 9.6 Product characteristics with a market constraint.

		Product Name	P	Q	R
		Marginal profit	8	4	5
Resource	**Type**	**Available**	**Required**	**Required**	**Required**
Blanking	<	1200	4	6	8
Drilling	<	1200	8	4	6
Grinding	<	1200	4	8	4
Heat treatment	<	1200	10	5	8
Limited demand	<	40	1	0	0

Table 9.6a   LP simplex solution for a market constraint.

Tableau—Initial

Basic	EQ	Z	x_1	x_2	x_3	x_4	x_5	x_6	x_7	x_8	RHS
Z		1	-8	-4	-5	0	0	0	0	0	0
x_4	1	0	4	6	8	1	0	0	0	0	1200
x_5	2	0	8	4	6	0	1	0	0	0	1200
x_6	3	0	4	8	4	0	0	1	0	0	1200
x_7	4	0	10	5	8	0	0	0	1	0	1200
x_8	5	0	1	0	0	0	0	0	0	1	40

Tableau—1

Basic	EQ	Z	x_1	x_2	x_3	x_4	x_5	x_6	x_7	x_8	RHS
Z	0	1	0	-4	-5	0	0	0	0	8	320
x_4	1	0	0	6	8	1	0	0	0	-4	1040
x_5	2	0	0	4	6	0	1	0	0	-8	880
x_6	3	0	0	8	4	0	0	1	0	-4	1040
x_7	4	0	0	5	8	0	0	0	1	-10	800
x_1	5	0	1	0	0	0	0	0	0	1	40

Tableau—2

Basic	EQ	Z	x_1	x_2	x_3	x_4	x_5	x_6	x_7	x_8	RHS
Z	0	1	0	-0.88	0	0	0	0	0.63	1.75	820
x_4	1	0	0	1	0	1	0	0	-1	6	240
x_5	2	0	0	0.25	0	0	1	0	-0.75	-0.5	280
x_6	3	0	0	5.5	0	0	0	1	-0.5	1	640
x_3	4	0	0	0.63	1	0	0	0	0.13	-1.25	100
x_1	5	0	1	0	0	0	0	0	0	1	40

Tableau—3

Basic	EQ	Z	x_1	x_2	x_3	x_4	x_5	x_6	x_7	x_8	RHS
Z	0	1	0	0	0	0	0	0.16	0.55	1.91	921.82
x_4	1	0	0	0	0	1	0	-0.18	-0.91	5.82	123.64
x_5	2	0	0	0	0	0	1	-0.05	-0.73	-0.55	250.91
x_2	3	0	0	1	0	0	0	0.18	-0.09	0.18	116.36
x_3	4	0	0	0	1	0	0	-0.11	0.18	-1.36	27.27
x_1	5	0	1	0	0	0	0	0	0	1	40

Table 9.7 Modeling a price discount.

Resource	Type	Product Name	P	P1	Q	R
		Product Value	8	7	4	5
Resource	Type	Available	Required	Required	Required	Required
Blanking	<	1200	4	4	6	8
Drilling	<	1200	8	8	4	6
Grinding	<	1200	4	4	8	4
Heat treat	<	1200	10	10	5	8
P	<	40	1	0	0	0
P1	<	40	0	1	0	0

Suppose the marketing department looks for orders and discovers that it can sell up to another 40 P if it offers a one-dollar discount on each piece. (See Goldratt and Cox 1992, 311–13, for a similar situation.) Formulate the problem as shown in Table 9.7.

$$
\begin{aligned}
Z - 8x_1 - 7x_2 - 4x_3 - 5x_4 &= 0 \\
4x_1 + 4x_2 + 6x_3 + 8x_4 + x_5 &= 1200 \text{ Blanking} \\
8x_1 + 8x_2 + 4x_3 + 6x_4 + x_6 &= 1200 \text{ Drilling} \\
4x_1 + 4x_2 + 8x_3 + 4x_4 + x_7 &= 1200 \text{ Grinding} \\
10x_1 + 10x_2 + 5x_3 + 8x_4 + x_8 &= 1200 \text{ Heat treatment} \\
x_1 + 0x_2 + 0x_3 + 0x_4 + x_9 &= 40 \text{ Demand for P} \\
0x_1 + x_2 + 0x_3 + 0x_4 + x_{10} &= 40 \text{ Demand for P1}
\end{aligned}
$$

The solution is to make 40 P (x_1) for sale at $8.00 profit, 20 P1 ($x_2$) for sale at $7.00 profit, and 120 Q ($x_3$) for a total of $940 profit. This is better than the $921.82 of the previous example. There are 240 minutes of excess capacity for blanking and drilling (x_5, x_6) and none for grinding or heat treatment (x_7, x_8). There are 20 units of excess demand for P1 (x_{10}). The shadow prices show the benefits of getting additional orders for P (x_9) or adding grinding (x_7) or heat treatment (x_8) capacity. This example shows that LP simplex can help decide whether to offer a product at a discount to get additional sales.

The Degenerate Solution

If the demand for P in Table 9.6 is set at 60, the solution in Table 9.8 results.

The problem is that there is a tie for the leaving basic variable in the second tableau. Since 75/0.625 is not *less than* 660/5.5 (both equal 120) the

Table 9.8   A degenerate solution for LP simplex.

Tableau—2

Basic	EQ	Z	x_1	x_2	x_3	x_4	x_5	x_6	x_7	x_8	RHS
Z	0	1	0	-0.88	0	0	0	0	0.63	1.75	855
x_4	1	0	0	1	0	1	0	0	-1	6	360
x_5	2	0	0	0.25	0	0	1	0	-0.75	-0.5	270
x_6	3	0	0	5.5	0	0	0	1	-0.5	1	660
x_3	4	0	0	0.63*	1	0	0	0	0.13	-1.25	75
x_1	5	0	1	0	0	0	0	0	0	1	60

* probably 0.625

Tableau—3

Basic	EQ	Z	x_1	x_2	x_3	x_4	x_5	x_6	x_7	x_8	RHS
Z	0	1	0	0	0	0	0	0.16	0.55	1.91	960
x_4	1	0	0	0	0	1	0	-0.18	-0.91	5.82	240
x_5	2	0	0	0	0	0	1	-0.05	-0.73	-0.55	240
x_2	3	0	0	1	0	0	0	0.18	-0.09	0.18	120
x_3	4	0	0	0	1	0	0	-0.11	0.18	-1.36	0
x_1	5	0	1	0	0	0	0	0	0	1	60

computer program selects x_6 as the leaving variable. The subsequent operations, however, reduce x_3 (product R), a basic variable, to zero. This is called a *degenerate basic variable* (Hillier and Lieberman 1980, 43).

The solution is still optimal and there is apparently no problem with the shadow price for x_7 (heat treatment). The $1.91 shadow price for x_8, demand for P, is *incorrect*. If the demand for P rises to 61, an additional P ($8.00) can be produced only if two units of Q (at $4.00) are foregone to release 10 minutes of heat treatment. The shadow price for x_8 should be zero under these conditions.

Degenerate solutions are not common and the authors do not provide any tie-breaking algorithm. This example suggests manual testing of the shadow prices (by examining the incremental profit that results from an increment in the resource) when the solution is degenerate.

Linear programming allows the factory to select the product mixture that yields the most profit and identifies the constraint(s) for that mixture. LP subjugates the mixture to factory constraints (capacity), market constraints (demand), and contractual constraints (mandatory customer requirements). It also allows modeling of many "what–if" situations. The shadow prices from simplex tableaus often point to the best opportunities for capacity improvement projects.

10

Program and Project Management

ompetition in the worldwide business industry is fiercer than ever. Companies encourage managers to identify and react to market conditions. Managers realize that firefighting is not effective so projects, assigned to key personnel, link improvement objectives to the root causes of chronic problems. These improvement projects can be large in scale and thus slow to implement. Since time is of the essence and the competition is not standing still, a new approach to project implementation is needed. This new approach is *critical chain project management*, as described by Goldratt (1997).

Critical chain has delivered proven results for Fairchild Semiconductor's Mountaintop plant, which completed an 8-inch (200-mm) semiconductor wafer processing plant in 13 months.[1] "Project Raptor" (named for the velociraptor dinosaur because it's fast, highly intelligent, works in teams, and eats its competitors) required not only erection of the factory and its cleanroom environment but also equipment installation and operator training. The factory's first steel beams were erected in February, 1996. The first wafers came out in February 1997, and they met yield standards on the first day (Murphy, Lauffer, and Levinson 1997). Furthermore:

> The industry norm for production ramp-ups ranges from 24 to 30 months. Project Raptor used synchronous flow manufacturing (SFM) (Murphy and Saxena 1997, Murphy and Saxena in Levinson 1998) and an aggressive inventory position to ramp to 100 percent capacity in less than six months (Murphy, Lauffer, and Levinson 1998).

The next section introduces critical chain in more detail.

CRITICAL CHAIN

According to Goldratt (1997), typical project management problems can come in the form of three separate issues:

1. The most common problem involves the project budget. Project financial support comes from cost estimates derived at the beginning of the project. If the project estimates are not 100% accurate, *financial overruns* occur.

2. The second issue involves *time overruns*. Project completion time estimates are often inaccurate. Delays in the schedule escalate as time passes.

3. The third problem involves *compromising content*. Poor quality usually occurs when time and/or money is in short supply.

All of these conditions stem from uncertainties found within all project management planning. These embedded uncertainties are the major causes of project mismanagement and delays in completing project goals.

Critical chain project management teaches the project leader to deal with all uncertainties within a project as follows. First, all tasks of a project do not receive equal priority. Unlike many conventional project management techniques, critical chain aligns with the *critical path method* (CPM) by identifying the longest chain of dependent activities. This becomes the critical chain of tasks. There is a tie-in with the theory of constraints here. The manufacturing constraint is the operation with no excess capacity, the critical chain is the *series of activities with no slack time*. The critical chain takes into account any task dependencies and contention for a given resource. Core team members can easily prioritize activities through the critical chain task schedule.

Tasks not on the critical chain are feeding tasks. In critical chain, these tasks do not receive the emphasis they would in conventional project management. The vital tasks of the critical chain receive focus. This focus eliminates the temptation to work on everything while accomplishing nothing. *The critical chain always takes priority*. Other tasks feeding the critical chain are in feeding chains. They begin a calculated time ahead of their interaction with the critical chain, also known as a feeding buffer, so they do not impact the completion of any critical chain tasks.

Tu Yü, a commentator on Sun Tzu's *Art of War*, described this concept hundreds of years ago. This is an example of how the life-and-death nature of military competition drives the development of management techniques that only later find their way into civilian enterprises.

Now those skilled in war must know when and where a battle will be fought. They measure the roads and fix the date. They divide the army and march in separate columns. Those who are distant start first, and those who are near by, later. Thus, the meeting of troops from a distance of [300 miles] takes place at the same time. It is like people coming to a city market (Sun Tzu 1963, 99).

This brief excerpt shows how:

1. The planners estimate activity completion times (measure the roads)

2. Activities take place in parallel (the army marches in separate columns)

3. The activities that require the most time start first, those that require less time start later

4. The activities converge smoothly to result in completion

This is the essence of project management.

The second major difference is the focus of the project leader on the performance of the project as measured by the different aspects of the critical chain, not individual task performance. Most projects begin with a list of tasks to accomplish toward a specific goal. Individuals responsible for the completion of tasks estimate completion times. To protect the individual task time upon which most people are measured, most estimates contain "safety" in the form of additional time to ensure completion by the date quoted. These estimates account for "uncertainties" that the individuals have experienced in the past. Most estimates contain at least 50% safety task time for each task quoted. Critical chain takes the safety margins for all the tasks, aggregates them, and moves the sum to the end of the project. This additional time, known as the *project buffer*, protects the completion date of the project for the critical chain of events.

From an organizational culture viewpoint, this change affects the way people think about project tasks. First, people realize they must start on a task immediately upon assignment. Removal of individual tasks' safety margins relocates the safety margin to the project buffer. If people have safety time built into a task they can delay the start until the last possible minute. Dr. Goldratt calls this the *student syndrome*, which is similar to a student waiting until the night before the due date to write a term paper.

The second cultural change is that people will not be able to work on several tasks at once. This is multitasking. Multitasking means working on several projects at once, and this often delays completion of the whole.

Critical chain prioritizes the activities so the important tasks are completed in the necessary order. Drivers for this change include (1) shortened task times due to removal of individual safety margins and (2) the measurements that are part of the critical chain application.

This transfer from conventional project management techniques shifts the focus from managing variation in individual task performance to managing the overall project. The two key measurements of project performance in critical chain are (1) the percentage of the critical chain completed and (2) the amount of the project buffer consumed. The relationship between these two measurements is the signal to management for appropriate action. Project review meetings focus on whether the progress of the critical chain is at a pace for completion without consuming the project buffer. The project buffer is broken down into three sections; notice the similarity to the zones in the manufacturing constraint buffer.

1. Consumption of less than one-third signals that the project is in the "OK zone" and should proceed as is.

2. If consumption falls between one-third and two-thirds of the project buffer, plans to recover upcoming critical chain tasks are necessary ("Watch and plan" zone).

3. If more than two-thirds of the project buffer is consumed, implementation of planned actions should happen at once ("Act" zone).

In this environment, the role of the project leader shifts from a focus on all tasks to tasks that are on the critical chain. Also, focus remains as necessary for any feeding chains that may be in danger of affecting the ability to start a task on the critical chain. Mountaintop uses a software package called ProChain from ProChain Solutions, Inc. for this analysis.

Critical Chain Planning

Planning for projects starts with the initial analysis within the core team. Resource managers and subject matter experts supply the information on tasks needed to complete the project. These tasks include resources from departments including manufacturing information systems (MIS), environmental and safety, human resources, facilities, product engineering, process engineering, manufacturing, purchasing, and planning.

Each department lists tasks necessary to completing their portion of the overall project. An initial project network develops from the individual tasks, and the overall project duration is forecasted from the project network. The project network includes resources required, task durations, and inter-task dependencies.

Projects can be singular or multiple depending on the amount of shared resources. Projects run as a single project environment have very little sharing of resources between them and other projects. Multi-project environments such as design of multiple new products must often share resources like access to manufacturing facilities, layout, and reliability testing.

The individual tasks load into an analysis software package called ProChain Plus. The software collects the information to create a modified Gantt chart with the critical chain identified. Summary tasks that consist of several individual tasks that comprise the bulk of the work determine progress. However, the major focus for all members of the core team remains on the individual tasks in the critical chain. *Improving (reducing) the duration of the critical chain is the only way to complete the project ahead of the scheduled finish date.* This again ties in with the theory of constraints for manufacturing, where elevation of the constraint is the only way to increase capacity.

If the initial project network provides an unsatisfactory completion date, individual task times can be modified. Individual tasks may contain unneeded safety or buffer times that can be removed. Another way to reduce project duration is by looking at a different way to complete the task or adding resources from another task. The initial critical chain for project completion is often too long. This is a frequent occurrence in the planning stage of the network. Eliminating task dependencies shortens task times and ultimately reduces the overall project length.

As an example, Project Raptor discarded some paradigms about task dependencies in bringing a new factory online:

1. Total completion of the building is not a prerequisite for equipment installation. Equipment installation can often occur as soon as floor space and utilities are available.

2. Equipment installation is not a prerequisite for operator training. Operators can be sent to the equipment vendor's site for training.

Several reports monitor the project for both resource managers and the project leader. The resource manager report allows the manager of each area to see what tasks the group must perform to complete the critical chain. The emphasis shifts from how much of a task is complete to how much remains. This is usually a much more accurate measure of real progress on a task. This is the appropriate information for updating the overall project.

The project leader report shows the status of the critical chain. Details show which tasks are currently in progress, how much of the critical chain is complete, and consumption percentage of the buffer. Again, the project leader focuses effort on completing critical chain tasks instead of every project task. Allocation of limited organizational resources improves effectiveness through this type of planning.

Resources for Critical Chain Project Management

The scope of projects can demand a wide range of talent and require relentless commitment to keep moving forward. Individuals supply the expertise to manage the project and form a core team. Individuals selected for core teams should have backgrounds in areas such as production, engineering, facilities, safety, human resources and project management.

Core team members fit into several different categories defined by the critical chain project management application. Descriptions of these roles and responsibilities follow.

1. The *project leader's* primary responsibility is the timely completion of the entire project. The leader ensures the completion of project tasks, on time and within budgetary constraints, by directing the actions of the core team members.

2. The *critical chain experts* possess the knowledge of the critical chain software. Their responsibilities include loading all tasks necessary for project completion. Reports from the information database inform the project leader and resource managers of critical chain task adherence.

3. *Resource managers* command the required subordinates to accomplish tasks within their resource area. These areas span a wide range of disciplines, all which are necessary for a project of this magnitude.

   • Resource groups report directly to the resource managers. These groups include the employees who will actually complete the specific task requirements.

Assessment

Determination of a project's success relies on three measurements:

1. The first measurement of success includes the two hardest variables to control in any project—project duration and financial budget. Speed is of the essence. The critical chain project management application monitors the adherence to the critical chain schedule and the amount of project buffer consumed. This measurement of time details the progress of the overall project and focuses the core team on necessary issues.

2. The second measurement of success is that of financial benefits. The critical chain initiates additional expenditures only when the overall project deadline is in jeopardy. This controls costs throughout the project.

3. The third and last measure of success is the potential to increase manufacturing capacity levels, thus increasing market share and return on investments (ROI).

The Importance of Time in Project Management

The previous section underscored a key concept: *speed is of the essence*. Planners often make the mistake of subordinating speed to cost containment. The famous Russian field marshal Aleksandr V. Suvorov (1729–1800) described the importance of speed more than 100 years before anyone ever heard of project management:

> Money is dear; human life is still dearer; but time is dearest of all. One minute decides the outcome of a battle, one hour the success of a campaign, one day the fate of empires (Menning 1986).

When market success depends on getting products to the market ahead of competitors, this statement is right on the mark. Suvorov subordinated money to time. Rapid victory in warfare usually reduced both its human and financial costs. The same lesson applies to marketing warfare.

The next sections provide overviews of some specific project management and assessment tools. The *critical path method* (CPM) ties in very closely with critical chain, and it is a valuable tool for identifying the critical chain. PERT (program evaluation and review technique) is a related method. *Present value analysis* is a managerial (not cost) accounting technique that accounts for the value of time. It discounts future cash flows according to the company's required rate of return. A dollar today is more valuable than a dollar next year, and a dollar next year is worth more than one two years from now. This is an argument for completing a revenue-generating project (like a factory) as quickly as possible instead of focusing on costs. Present value analysis, despite its analytical power and scientific basis, does not account for intangibles such as being first (or second) to the marketplace. An in-depth discussion of these techniques is beyond the scope of this book but the reader should become familiar with the concepts.

CRITICAL PATH METHOD

The Gantt chart is a simple and easily understood project management tool. It can incorporate simple precedence relationships (task dependencies) but not complicated ones. PERT and CPM work with complicated task networks, and computer algorithms are available for them.

J. E. Kelly of Remington Rand and M. R. Walker of DuPont developed CPM in 1957. The Special Projects Office of the U.S. Navy worked with Booz, Allen, and Hamilton to develop PERT in 1958 for the Polaris missile program. Polaris involved thousands of contractors, which should give an idea of the scope and complexity of this project. PERT was credited with shortening the project by 18 months (Heizer and Render 1991, 698–699). The authors continue with the procedure for PERT or CPM:

1. Define the project and identify all the significant tasks or activities.

2. Identify precedence relationships between activities. Which tasks are prerequisites for other tasks?

3. Draw a network to connect all the tasks or activities.

4. Estimate the time and cost for each task.

5. Find the longest path through the network, the path that has no slack time. *This is the critical path.*

6. The network now becomes a tool for planning, scheduling, monitoring, and controlling the project.

This is really no different than critical chain. A couple of terms deserve explanation:

1. Events are related to tasks and activities.

    • An *event* is illustrated in the project network by a circle (node). It has a starting time and a completion time.

    • An *activity* is shown by an arrow. It represents something that happens over time, such as a task.

2. *Slack time* is like excess capacity in a factory. The difference between a task's earliest and latest (consistent with completing the project in the minimum possible time) starting times is its slack time. The tasks that have no slack time—tasks that must start as early as possible to achieve on-time completion—constitute the critical path.

Table 10.1 and Figure 10.1 provide an example of CPM.

Table 10.1 Project events.

Event	Prerequisite Event(s)	Activity	Time Required
1 (start)			—
2	1	1–2	4
3	2	2–3	6
4	1	1–4	3
5	2,4	2–5	0 (dummy)
		4–5	3
6	4	4–6	4
7	5,6	5–7	0 (dummy)
		6–7	2
8 (finish)	3,7	3–8	0 (dummy)
		7–8	2

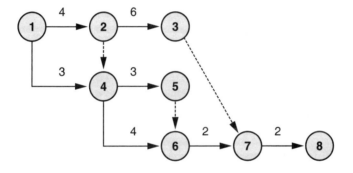

Figure 10.1 Project network.

The next step is to draw the project network. The rules for doing this are (Hillier and Lieberman 1980, 247):

1. Each *branch* or *activity* represents a task

2. Each node represents an *event*, the completion of all activities that lead into that node

   - The event must be completed before any activity that leads out of it can begin. The event is a prerequisite for the subsequent activities.

3. No more than one branch or activity may connect any two nodes

- Dummy activities, which are represented by dashed lines and which have zero completion times, show precedence relationships only

Here is an example of a dummy activity. Events 2 and 4 are prerequisites for event 5. The dashed line between 2 and 4 shows that the activity that leads to 5 can't begin until both 2 and 4 are complete.

Table 10.2 shows how to identify the *earliest possible starting and finishing times* by working *forward* through the network.

This shows that the project can be completed in 12 units of time. Now identify the *latest possible finishing times* by working *backward* through the network (Table 10.3). Begin with event 8, which can be completed at $t = 12$. The next row shows that event 7 must finish at $t = 10$ if 8 is to finish at $t = 12$. If 7 is to finish at $t = 10$, 6 must finish at $t = 8$. Also, because 5 is a prerequisite for 7, activity 6–7 cannot take place without completion of 5. If 7 is to finish at $t = 10$, 5 (like 6) must finish at $t = 8$. Continue to work backward through the table to get the latest allowable finishing time for each event.

Finally, the slack for each activity is the difference between the latest allowable completion time for an event minus (earliest possible start + activity time) (Table 10.4).

Table 10.2 Calculation of earliest possible finishing times.

Event	Prerequisite Event(s)	Activity, Time Required	Earliest Start + Activity	Maximum = Earliest Finish
1 (start)		—	0	
2	1	1–2 4	0 + 4	4
3	2	2–3 6	4 + 6	10
4	1	1–4 3	0 + 3	3
5	2	4–5 3	4 + 3	7 (greater of
	4		3 + 3	7 and 6)
6	4	4–6 4	3 + 4	7
7	5	6–7 2	7 + 2	9
	6		7 + 2	
8 (finish)	3	7–8 2	10 + 2	12 (greater of
	7		9 + 2	12 and 11)

Table 10.3 Calculation of latest allowable finishing times.

Event	Dependent Event(s)	Activity, Time Required	Latest Finish– Activity	Minimum = Latest Finish
8 (finish)	—	(finish)	—	12
7	8	7–8 2	12 – 2	10
6	7	6–7 2	10 – 2	8
5	7	6–7 2	10 – 2	8
4	5	4–5 3	8 – 3	4 (lesser of 5 and 4)
	6	4–6 4	8 – 4	
3	8	7–8 2	12 – 2	10
2	3	2–3 6	10 – 6	4 (lesser of 4 and 5)
	5	4–5 3	8 – 3	
1	2	1–2 4	4 – 4	0 (lesser of 0 and 1)
	4	1–4 3	4 – 3	

Table 10.4 Calculation of slack times.

Event	Latest Finish	Earliest Start	Activity, Time	Slack
1 (start)	0	0	—	0
2	4	0	1–2 4	0
3	10	4	2–3 6	0
4	4	0	1–4 3	1
5	8	3	4–5 3	2
6	8	3	4–6 4	1
7	10	7	6–7 2	1
8 (finish)	12	10	7–8 2	0

The critical path is therefore 1–2–3–8. Any delay in this chain of events will delay the whole project. Figure 10.2 is Figure 10.1 with earliest possible and latest acceptable completion times. The two quantities match on the critical path.

Crashing (Acceleration) of Critical Path Activities

This book has already shown why productivity improvement techniques like total productive maintenance and single-minute exchange of die do not increase factory capacity except at the constraint. The same principle

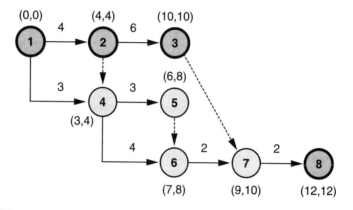

Figure 10.2 Project network, earliest possible and latest allowable completion times.

applies to critical chain project management. It is sometimes possible to perform critical path activities on a *crash basis*. This means spending extra money to expedite the activities. There is obviously no benefit to expediting an activity that is not on the critical path.

Acceleration of critical path activities can create a new critical path. This is similar to an action that breaks a manufacturing constraint (by increasing its capacity) and makes another operation the constraint.

Program Evaluation and Review Technique (PERT)

PERT differs from CPM in its ability to account for *uncertainty* in activity times. CPM assumes that activity times are deterministic; they can be stated exactly. PERT uses an expected time, t_e, and a variance, σ^2, where

$$t_e = \frac{a + 4m + b}{6} \qquad \sigma^2 = \left(\frac{1}{6}(b-a)^2\right)$$

and a is the optimistic estimate of the time, m is the *most likely* (not average) time, and b is the pessimistic estimate. Note the weighting of the most likely time (m) in calculation of the expected time. The underlying statistical distribution could be a beta distribution but a normal distribution is often assumed. Per Hillier and Lieberman (1980, 253), PERT uses the following assumptions:

1. Activity times are assumed to be *statistically independent*.

2. The critical path (as determined by the expected times) *always* requires more time than any other path.

    *   The sum of the independent times in the critical path is therefore the sum of the expected times. The variance of the project completion time is the sum of the activities' variances.

3. The project time will follow a normal distribution. Even if the activity times on the critical path do not follow the normal distribution, the central limit theorem suggests that their sum will be approximately normal.

This allows calculation of the probability that the project will finish in less than a given amount of time.

A project's critical path is closely related to the manufacturing constraint. The concept of slack (surplus capacity) from linear programming makes this relationship even clearer through its tie-in with slack time in a project. The next section treats present value analysis, a major engineering economic analysis technique for project assessment.

PRESENT VALUE ANALYSIS

Present value analysis assesses projects against a required rate of return, i. It accounts for the *time value of money* by discounting future cash flows. It ties in with CPM by allowing the user to determine the desirability of crashing an activity on the critical path. It can answer questions like, "Is it worth an extra \$100,000 to crash an activity on the critical path to get a new factory online three months sooner?" Here are the basics:

1. *The net present value of any combination of incomes and outlays is simply the sum of their present values.*

    *   The following equation discounts all future transactions (F_k, an outlay or a receipt at the end of the kth period) to the present when the required rate of return on investments is i (expressed as a fraction, $8\% = 0.08$).

    $$NPV = \sum_{k=0}^{N} F_k \frac{1}{(1+i)^k}$$

    where the project or activity runs from the present ($k = 0$) through N periods. There are, however, more convenient equations for summarizing annual or other periodic payments.

- The required rate of return (also known as the hurdle rate or the opportunity cost of capital) is generally what the organization thinks it should be earning on its investments. For individuals, the criterion might involve interest on debt. Suppose someone who is paying eight percent interest on a mortgage has $10,000 to invest. If they can earn ten percent in the stock market, that investment is favorable. If they can earn only six percent in the stock market, they are better off using the money to reduce the mortgage's principal.

2. If the project or activity has a positive net present value, it is favorable under the required rate of return.

Time Value of Money

As a simple example, suppose that one can earn eight percent interest (compounded annually for simplicity, methods are available for monthly or even continuous compounding) on investments. $100 today is worth $108 next year. In two years, it will be worth $108 + 0.08 × $108, or $108 × 1.08 = $116.64. This is the future value of a present investment. Alternately, a promised payment of $116.64 two years from today is worth $100 to this investor. There are, of course, mathematical formulas for dealing with different kinds of income streams (Table 10.5), see also Riggs (1977, 164–68).

The factor symbols can be read as follows. The sinking fund factor, $(A/F,i,N)$, is shorthand for "annual payment A required to deliver a future payment of F, given interest rate i, after N years." Years can be replaced by some other period. For example, 12 percent annual interest compounded monthly can be handled by letting $i = 0.01$ and using months instead of years.

This kind of information allows some very intelligent and informed decisions. Suppose that a company expects to earn 20 percent on its money. A $90,000 investment in a new machine will elevate a manufacturing constraint and the additional capacity will earn $30,000 per year for six years. Here are some answers one might get from less sophisticated decision-making processes:

- Cost accounting: "The equipment we have is fully depreciated. If we buy the new equipment, depreciation will impact the bottom line."

- Payback method (a very conservative, risk-averse criterion): "The payback period is three years, the current standard for new investments is two years."

$$\text{Payback} = \frac{\text{Initial outlay or investment (dollars)}}{\text{Benefits}\dfrac{\text{dollars}}{\text{year}}} = \text{years,}$$

and it answers the question, "How long will the investment take to pay for itself?"

Figure 10.3 shows the revenue stream for the investment. (It is often useful to draw a picture like this.)

Table 10.5 Time value of money.

Factor	Symbol and Formula	Timeline
Compound amount, single payment	$(F/P,i,N) = (1+i)^{N}$	P ... F 0 1 2 N Time
Present worth, single payment	$(P/F,i,N) = \dfrac{1}{(1+i)^{N}}$	
Compound amount, uniform series	$(F/A,i,N) = \dfrac{(1+i)^{N}-1}{i}$	A A A A F 0 1 2 N Time
Sinking fund	$(A/F,i,N) = \dfrac{i}{(1+i)^{N}-1}$	
Present worth, uniform series	$(P/A,i,N) = \dfrac{(i+1)^{N}-1}{i(i+1)^{N}}$	P A A A A 0 1 2 N Time
Capital recovery	$(A/P,i,N) = \dfrac{i(i+1)^{N}}{(i+1)^{N}-1}$	

A = annual payment that takes place at the *end* of the year
F = payment that takes place at the *end* of N years
i = rate of return, hurdle rate, or interest rate, expressed as a fraction
N = number of years
P = payment that takes place now (*beginning* of the first year, $t = 0$)

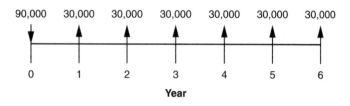

Figure 10.3   Revenue timeline for present worth analysis.

The net present value is NPV = -$90,000 + $30,000(P/A,0.20,6), where (P/A,0.20,6) equals 3.3255. The net present value is therefore $9765.30 and, since this is greater than zero, the investment fulfills the company's requirement for rate of return. Another way to state this is, "$30,000 per year for six years is, with a 20% required rate of return, the same as having $99,765.30 today." Since $99,765.30 is more than the $90,000 that the company must spend to get it, the investment is favorable. Present value analysis would approve this investment where "bean counting" and other decision-making tools would reject it. This underscores two key principles for entrepreneurial success:

1. Errors of omission (doing nothing), not errors of commission (doing the wrong thing) are the predominant causes of failure in most enterprises.

2. The organization, specifically through its performance measurement systems and decision tools, is often its own worst enemy. "With tools like cost accounting to run the factory, who needs competitors?"

   - The payback criterion has enough respectability to receive mention in engineering and managerial economics books but these sources emphasize its deficiencies. It ignores both the time value of money and benefits that take place after the payback period (Riggs 1977, 393).

Rate of Return

Rate of return (ROR) is intuitively attractive for two reasons:

1. There is no need to state a required rate of return

2. It *looks* like it should allow side-by-side comparison of different investment options

ROR is difficult to calculate by hand but it is straightforward by computer. The idea is to find the rate of return, i, that makes the net present value of all payments and receipts zero. Bisection, the Newton-Raphson method, and similar approaches adjust i in

$$NPV = \sum_{k=0}^{N} F_k \frac{1}{(1+i)^k}$$

(or an equation that uses the factors in Table 10.5) until $NPV = 0$.

ROR project comparisons usually but not always yield the same results as NPV project comparisons. A deficiency of ROR is that it assumes there are opportunities to reinvest funds that are recovered during the project's life at the calculated ROR, and this may not be true (Riggs 1977, 274).

Fairchild Semiconductor's Mountaintop plant proved the effectiveness of the critical chain method by completing a new semiconductor factory in record time. Related techniques include CPM and PERT, and there is a connection between these and the theory of constraints. The manufacturing constraint is the operation with no slack (excess) capacity, and the critical path is the set of activities for which there is no slack time.

Present value analysis aids project selection, and it can also quantify the effects of crashing (paying more money to accelerate) an activity in a project's critical path.

ENDNOTE

1. The diameter specification is metric, and 200 mm is actually slightly smaller than 8 inches (203 mm).

11

Conclusion

This book has presented lean enterprise as a set of synergistic and mutually supporting techniques and programs. All focus on the elimination of *friction*, or non-value-adding activities, from the enterprise.

The concept of friction (Japan's muda, or waste) is very simple but everyone in the organization must remember that friction is easy to overlook. The ability to identify friction on sight may well have been Henry Ford's principal success secret, and his achievements and bottom line speak for themselves. The examples in this book have hopefully equipped the reader with the same skill plus the ability to teach others.

CULTURAL TRANSFORMATION

The implementation of lean manufacturing requires *cultural transformation*, a change in "the way we do things around here." There are three major prerequisites: management commitment, job security, and elimination of rigid job classifications. Management commitment is a prerequisite for any quality or productivity improvement initiative. Job security and worker flexibility go together because job classifications are labor's defense against layoffs.

Everyone in the organization must understand and buy into the lean concept. Change agents can present lean manufacturing as the ultimate defense against the lure of cheap offshore labor. The American origins of lean manufacturing offer a special advantage in American workplaces, while the unquestionable prosperity of lean practitioners is a selling point in any organization.

SINGLE-UNIT FLOW

The importance of single-unit, or at least small-batch, production flow is a vital point. Batching and queuing increase cycle time by forcing units to wait. Even one batch-and-queue operation can generate inventory despite the best just-in-time production control system. Batching also complicates statistical process control and process capability assessment. Lean manufacturing techniques like single-minute exchange of die help reduce batch sizes.

ANOTHER FORD SUCCESS SECRET: SUPPRESSION OF VARIATION IN PROCESSING TIME

Variation in processing times creates a hurry-up-and-wait effect that generates inventory even in systems with excess capacity. Henry Ford's statement that each worker should have all the time he needs but not a second more sounds like a reckless ambition to run a balanced factory at close to 100 percent capacity! The fact that Ford's system *worked*, plus the concept of takt time and equation 4.1 (p. 70), offers a priceless lesson. The only way to achieve almost 100 percent utilization (without piling up inventory and waiting time) is to suppress variation in processing time and work arrival time.

PRODUCTION MANAGEMENT AND THE THEORY OF CONSTRAINTS

The reader should have gained an understanding of Dr. Eliyahu Goldratt's theory of constraints, synchronous flow manufacturing, and the drum-buffer-rope production management system. SFM is simply another form of just-in-time production management but it offers a couple of advantages over kanban systems. First, the inventory buffer at the constraint prevents lost production due to stoppages at non-constraint operations. Second, the DBR system does not require production control links between all adjacent operations. The link between the constraint and production starts is usually sufficient.

The chapter on linear programming should have provided an appreciation of a powerful analytical tool that is a perfect complement to TOC. Linear programming allows its user to select the optimum product mix subject to a wide variety of constraints. These include not only factory capacity (the focus of TOC) but also contractual requirements and market constraints. It allows scientific modeling of "what–if" situations, such as elevating the constraint, changing the product so it requires less of a certain resource, or offering discounts to increase sales.

Critical chain project management is an offshoot of the theory of constraints. Under TOC, the capacity-constraining resource has no slack capacity and it limits production. Project activities on the *critical path* have no slack time, and they dictate the minimum time in which a project can be completed. Factories can improve their capacities by elevating their constraints, and project managers can shorten completion times by crashing (accelerating, often by providing more resources) activities in the critical path.

SUPPLY CHAIN MANAGEMENT

Lean enterprise requires the entire supply chain to adopt lean culture and lean techniques. Large batch deliveries by suppliers and batch processing by subcontractors can add inventory and cycle time to the leanest factory's operations. Supplier development is a vital concept, and customers sometimes need education as well.

Good communications are essential to supply chain management and there is no excuse for their absence. The Ford Motor Company relied on telephones to run an advanced freight management system during the 1920s. Today we have communication satellites, the Internet, and corporate extranets. They allow linkage, for example, of production control and transportation systems.

Long geographical distances make JIT deliveries harder in the United States than in Europe or Japan, but FMSs and third party logistics systems overcome some of the difficulties. Russia will need to appreciate this as it industrializes.

A SUMMARY OF THE LEAN MINDSET

The continuous improvement (kaizen) mindset was central to the early Ford Motor Company's organizational culture, and it now plays the same role in leading Japanese firms. This thought process requires us to never take any part of any job for granted; anyone who wants to see friction that everybody else overlooks must adopt it. Never assume that, because "we've always done it that way," there is not a better way to do it. The standard is only today's "one best way," and a better best way can and should supersede it tomorrow.

References

Aitken, Hugh G. J. 1960. *Scientific Management in Action: Taylorism at Watertown Arsenal*, 1908–1915. Princeton: Princeton University Press.

Alvarado, Rudolph, and Sonya Alvarado. 2001. *Drawing Conclusions on Henry Ford: A Biographical History through Cartoons*. Ann Arbor: University of Michigan Press.

Arnold, Horace Lucien, and Fay Leone Faurote. 1915. *Ford Methods and the Ford Shops*. New York: The Engineering Magazine Company. Reprinted in 1998, North Stratford, NH: Ayer Company Publishers Inc.

Australia Fights Methane. 2001. *Chemical and Engineering News* (June 18, 2001): 104.

Bakerjian, Ramon, ed. 1993. *Tool and Manufacturing Engineers Handbook: Volume 7, Continuous Improvement*. Dearborn, MI: Society of Manufacturing Engineers.

Basset, William. 1919. *When the Workmen Help You Manage*. New York: The Century Company.

Bennett, Harry, as told to Paul Marcus. 1951. *Ford: We Never Called Him Henry*. New York: Tom Doherty Associates, Inc.

Briscoe, Scott. 2001. The Tiger: Poised to Strike. *APICS—The Performance Advantage* (April 2001): 40–44.

Caravaggio, Michael. 1998. Total Productive Maintenance. In *Leading the Way to Competitive Excellence: The Harris Mountaintop Case Study*. Levinson, William, ed. Milwaukee, WI: ASQ Quality Press.

Carter, Todd. 2001. Driving Costs Down. *APICS—The Performance Advantage*, April 2001, 47–50.

Clausewitz, Carl von (d. 1831). 1976 (M. Howard and P. Paret translation). *On War*. Princeton, NJ: Princeton University Press.

Courtney, Steven. 1992. Modern Manufacturing in an Old-Fashioned Setting. *Hartford Courant* (December 7, 1992).

Cox, James F. III, and John H. Blackstone, Jr. 1998. *APICS Dictionary*, 9th ed. Alexandria, VA: APICS—The Educational Society for Resource Management (APICS was formerly the American Production and Inventory Control Society).

Crow, Carl. 1943. *The Great American Customer*. New York: Editions for the Armed Services, Inc.

Cubberly, William H. and Ramon Bakerjian, eds. 1989. *Tool and Manufacturing Engineers Handbook, Desk Edition*. Dearborn, MI: Society of Manufacturing Engineers.

Dieter, George. 1983. *Engineering Design: A Materials and Processing Approach*. New York: McGraw-Hill.

Feigenbaum, Armand. 1991. *Total Quality Control*, 3rd ed. Milwaukee: ASQ Quality Press.

Flanders, R. E. 1925. Design Manufacture and Production Control of a Standard Machine. *Transactions of ASME*, 46 (1925).

Ford, Henry, and Samuel Crowther. 1922. *My Life and Work*. New York: Doubleday, Page & Company.

———. 1926. *Today and Tomorrow*. New York: Doubleday, Page & Company (Reprint available from Productivity Press, 1988).

———. 1930. *Moving Forward*. New York: Doubleday, Doran & Company.

Gardner, Daniel. 2001. Movers and Shapers: The Impact of Logistics on Global Supply Chains. *APICS— The Performance Advantage* (May 2001): 29–33.

Gardner, Les, and Frank Nappi. 2001. The Total Impact of Minor Stoppages. "The 6th Annual Lean Management and TPM Conference," sponsored by Productivity Inc. October 25–26, 2001, Dearborn, MI.

Gilbreth, Frank. 1911. *Motion Study*. New York: D. Van Nostrand Reinhold.

Goldratt, Eliyahu. 1997. *Critical Chain*. Croton-on-Hudson, NY: North River Press.

Goldratt, Eliyahu, and Jeff Cox. 1992. *The Goal*. Croton-on-Hudson, NY: North River Press.

Goldratt, Eliyahu, and Robert E. Fox. 1986. *The Race*. Croton-on-Hudson, NY: North River Press.

Halpin, James F. 1966. *Zero Defects*. New York: McGraw-Hill.

Harry, Mikel, and Richard Schroeder. 2000. *Six Sigma: The Breakthrough Management Strategy Revolutionizing the World's Top Corporations*. New York: Currency Doubleday.

Heizer, Jay, and Barry Render. 1991. *Production and Operations Management*, 2nd ed. Needham Heights, MA: Allyn and Bacon.

Hillier, Frederick S., and Gerald J. Lieberman. 1980. *Introduction to Operations Research*. Oakland, CA: Holden-Day.

Holt, James R., and Scott D. Button. Sharing the Destiny across Multiple Business Units: The Supply Chain Solution. *SIG Synergy, APICS* (September 2000).

Hradesky, John. 1995. *Total Quality Management Handbook*. New York: McGraw-Hill.

Imai, Masaaki. 1997. *Gemba Kaizen*. New York: McGraw-Hill.

Jacques, March Laree. 2001. Big League Quality: Deming ways change 115-year-old Louisville Slugger manufacturer. *Quality Progress* (August 2001): 27–34.

Juran, Joseph. 1992. *Juran on Quality by Design*. New York: The Free Press.

Juran, Joseph M., and Frank Gryna. 1988. *Juran's Quality Control Handbook*, 4th ed. New York: McGraw-Hill.

Kalpakjian, Serope. 1984. *Manufacturing Processes for Engineering Materials*. Reading, MA: Addison-Wesley.

Kastelic, F. M. 1993. Exporting is easy. *Manufacturing Engineering* (August 1993): 104.

Kim, Irene. 2001. Powering Down: Process Companies Explore Innovative Technology- and Market-Based Solutions to the Energy Crisis. *Chemical Engineering Progress* (August 2001): 10–12.

Laraia, Anthony C., Patricia E. Moody, and Robert W. Hall. 1999. *The Kaizen Blitz*. New York: John Wiley & Sons.

Larson, Melissa. 1998. Ergonomic Workstations Boost Productivity. *Quality* (March 1998): 44–47.

Lathin, Drew, and Ron Mitchell. 2001. Learning from Mistakes: For Lean Manufacturing to Work, You Must Integrate the Social and Technical. *Quality Progress* (June 2001): 39–45.

Levinson, William. 1994. *The Way of Strategy*. Milwaukee: ASQ Quality Press. Now available from iUniverse.com .

———, ed. 1998. *Leading the Way to Competitive Excellence: The Harris Mountaintop Case Study*. Milwaukee, WI: ASQ Quality Press.

———. 2000. *ISO 9000 at the Front Line*. Milwaukee: ASQ Quality Press.

———. 2000a. SPC for Real-World Processes: What to Do When the Normality Assumption Doesn't Work. Paper presented at the ASQ's annual quality conference, Indianapolis.

———. 2002. *Henry Ford's Lean Vision: The Enduring Waste-Free Principles from the First Ford Motor Plant*. Portland, OR: Productivity Press.

Levinson, William, and Frank Tumbelty. 1997. *SPC Essentials and Productivity Improvement: A Manufacturing Approach*. Milwaukee: ASQ Quality Press.

Liker, Jeffrey K., ed. 1998. *Becoming Lean: Inside Stories of U.S. Manufacturers*. Portland, OR: Productivity Press.

Mahan, Alfred Thayer. 1980. *The Influence of Sea Power Upon History, 1660–1805*. London: Bison Books.

Marciano, Michael. 1999. How Did Hartford Get into This Mess? *The Hartford Advocate*, http://www.hartfordadvocate.com/articles/hfdmess.html (as of 4/26/01).

Mege, Claude Jean. 2000. Is There a HAM in Your Future? *Manufacturing Engineering* (July 2000): 114–24.

Menning, Bruce W. 1986. Train Hard, Fight Easy: The Legacy of A. V. Suvorov and His 'Art of Victory.' *Air University Review* (December 1986): 79–88.

Moore, Albert W. 1996. 'Engine' Charlie was Right in 1952—and Still Is. *Manufacturing Engineering* (August 1996): 296.

Murphy, Robert, Jeff Lauffer, and William Levinson. 1997. Project Raptor. *Future Fab International* 1, no. 3: 117–20.

————. 1998. Velociraptor: Swift Predator, *Future Fab International* 1, no. 5: 117–23.

Murphy, Robert, and Puneet Saxena. 1997. Breaking Paradigms with Synchronous Flow Manufacturing. *Semiconductor International* (October 1997): 149–54.

————. 1998. Synchronous Flow Manufacturing. In *Leading the Way to Competitive Excellence: The Harris Mountaintop Case Study*. Levinson, William, ed. Milwaukee, WI: ASQ Quality Press.

Norwood, Edwin P. 1931. *Ford: Men and Methods*. Garden City, NY: Doubleday, Doran & Company Inc.

Ohno, Taiichi. 1988. *Toyota Production System*. Portland OR: Productivity Press.

Olexa, Russ. 2001. Pushing the Productivity Envelope. *Manufacturing Engineering* (May 2001): 72–84.

Peters, Thomas. 1987. *Thriving on Chaos*. New York: Harper & Row.

————. 1989. When surviving is not enough. Presentation to the Cornell Society of Engineers, April 28, Ithaca, NY.

Rerick, Ray, and Greg Klusewitz. 1996. Constraint Management through the Drum-Buffer–Rope System. SEMI/IEEE Advanced Semiconductor Manufacturing Conference and Workshop, November 12–14, 1996, Cambridge, MA.

"Resin Crisis Averted." 1993. *Semiconductor International* (October 1993): 26.

Richards, Bill. Inside Story. *Wall Street Journal* (June 17, 1996): R23.

Riggs, James L. 1977. *Engineering Economics*. New York: McGraw-Hill.

Robinson, Alan, ed. 1990. *Modern Approaches to Manufacturing Improvement: The Shingo System*. Portland: Productivity Press.

Sands, Allen. 1998. Customer Contact Teams. In *Leading the Way to Competitive Excellence: The Harris Mountaintop Case Study*. Levinson, William, ed. Milwaukee, WI: ASQ Quality Press.

————. 1998. Zero Scrap Actions. In *Leading the Way to Competitive Excellence: The Harris Mountaintop Case Study*. Levinson, William, ed. Milwaukee, WI: ASQ Quality Press.

Schonberger, Richard J. 1982. *Japanese Manufacturing Techniques: Nine Hidden Lessons in Simplicity*. New York: The Free Press.

————. 1986. *World Class Manufacturing*. New York: The Free Press.

Schragenheim, Eli. TOC and Supply Chain Management. *SIG Synergy, APICS* (September 2000). (APICS is the former American Production and Inventory Control Society, now The Educational Society for Resource Management)

Schragenheim, Eli, and H. William Dettmer. 2001. Constraints & JIT: Not Necessarily Cutthroat Enemies. *APICS—The Performance Advantage* (April 2001): 57–60.

Shingo, Shigeo. 1986. *Zero Quality Control: Source Inspection and the Poka-Yoke System*. Portland, OR: Productivity Press.

Sienko, Michell J., and Robert A. Plane. 1974. *Chemical Principles and Properties*, 2nd ed. New York: McGraw-Hill.

Sinclair, Upton. 1937. *The Flivver King*. Second printing, 1987. Chicago: Charles H. Kerr Publishing Company.

Smith, Catherine F. 1998. The Rhetoric of Record-Keeping II: Shorthand Writing. http://web.syr.edu/~cfsmith/congress/episodes/1789/comments/US/shorthand. html as of 8/21/01. An adaptation of Smith, C. F. (1994). Documenting Democracy in the First Federal Congress of the United States. Presented on panel *Rhetoric of Nation-Building*, Conference on College Composition and Communication.

Smith, Wayne. 1998. *Time Out: Using Visible Pull Systems to Drive Process Improvements*. New York: John Wiley & Sons.

Sorensen, Charles E. 1956. *My Forty Years with Ford*. New York: W. W. Norton & Company Inc.

Standard, Charles, and Dale Davis. 1999. *Running Today's Factory: A Proven Strategy for Lean Manufacturing*. Cincinnati, OH, Hanser Gardner Publications.

Steuben, Baron von. 1779. *Regulations for the Order and Discipline of the Troops of the United States*. http://www.2nc.org/steubman.htm .

Stuelpnagel, T. R. 1993. Déjà Vu: TQM Returns to Detroit and Elsewhere. *Quality Progress* (September): 91–95.

Sun Tzu (trans. by Samuel Griffith). 1963. *The Art of War*. New York: Oxford University Press.

Suzaki, Kyoshi. 1987. *The New Manufacturing Challenge*. New York: The Free Press.

The System Company. 1911. *How Scientific Management Is Applied*. 2nd revised edition. London: A. W. Shaw Company, Ltd.

———. 1911a. *How to Get More Out of Your Factory*. London: A. W. Shaw Company, Ltd.

Taylor, Frederick Winslow. 1911. *The Principles of Scientific Management*. New York: Harper Brothers. 1998 republication by Dover Publications, Inc., Mineola, NY.

———. 1911a. *Shop Management*. New York: Harper & Brothers.

Tsouras, Peter G. 1992. *Warrior's Words: A Dictionary of Military Quotations*. London: Arms and Armour Press.

Tsutsui, William M. 1998. *Manufacturing Ideology: Scientific Management in Twentieth-Century Japan*. Princeton: Princeton University Press.

Voiland, Douglas E. 2001. A Nice Problem to Have: A Commonsense Approach to TOC Can Save Even the Most Successful Company. *APICS—The Performance Advantage* (July 2001): 29–31.

Walker, Bill. 2001. Supply Chain Management. APICS meeting, Pittston, PA, March 14, 2001.

Walker, William. 2001a. Synchronized for Growth. *APICS—The Performance Advantage* (April 2001): 26–29.

Ward, John R. 1999. The Little Ships That Could. *American Heritage of Invention and Technology* (Fall 1999): 34–40.

Wentz, Martin. 1998. Teaming to Win. In *Leading the Way to Competitive Excellence: The Harris Mountaintop Case Study*. Levinson, William, ed. Milwaukee, WI: ASQ Quality Press.

Western Electric Co., Inc. 1956. *Statistical Quality Control Handbook*. Charlotte, NC: Delmar Printing Company.

Womack, James P., and Jones, Daniel T. 1996. *Lean Thinking*. New York: Simon & Schuster.

Index